TIDES OF HISTORY

TIDES OF HISTORY

A Review of Three Historic Tides
that Have Helped Shape the 20th Century:
Islam
Russia
Science

Louis Lavoie

1997

Published by:
Louis Lavoie
2675 Larch Lane
Plymouth, MN 55441

Library of Congress Catalog Card Number 96-95301

ISBN 1-57502-385-7

Printed in the USA by

MORRIS PUBLISHING

3212 E. Hwy 30
Kearney, NE 68847
800-650-7888

To my son

Chris Lavoie

Someone special

Contents

Figures & Maps

Figures Page

Maps

Table

Acknowledgements

It is always a special pleasure to acknowledge the help one has received in any creative enterprise. Those mentioned here will never truly know how grateful I am and how important their contributions have been. The staffs and students at Anoka/Ramsey and Lakewood Community Colleges, the Science Museum of Minnesota, Camp Unistar, and the Elderhostels where I first tried out my ideas were very supportive and constructive. Alan Willis at the University of Minnesota Cartography Laboratory was particularly helpful with the map preparations. I wish especially to note Pat Lockyear at Lakewood and Karen Engler, née Laun, at the Science Museum. No less than three senior vice presidents at Honeywell Inc. offered their protection when some of my ideas decades ago were not popular within the defense establishment. They know who they are; thanks. My family has been profoundly influential in this enterprise; thanks to my father Louis Sr., my mother Mary, my sister Harriet Showah, and my son Chris. I am deeply grateful for Connie Lavoie's special editorial help. My friends Roger Berg and Galen Sneesby were kind enough to read parts of the manuscript and offer excellent suggestions as was Ann Weiblen. My very good friend Professor Paul Weiblen at the University of Minnesota was especially helpful with Soviet science. Professor John Woods at the University of Chicago was very helpful in clarifying some issues in Islāmic history and the Arabic language. And finally, I wish I could thank the late John Beckmann for all he did to encourage and inspire me; this book would not have been written had he not been there.

Preface

This book is about the tides of history that define our time, the processes and events of such significance and importance that no "History" written in some future century would be able to ignore. This is not primarily a book about current events, though current events are included. The flow of current events without temporal context is bewildering if indeed not meaningless. The context is the tide of history, that which inescapably includes what has come before. Accordingly, a great deal of this book will selectively present what has come before so that sufficient context will be available to make some sense out of the present and give some hint of the future.

The selection of themes is not definitive. A lot is going on in areas other than the religious, political, and scientific, and even within these areas much has been omitted. Nevertheless, no general "History" of the 20th century will ever be written without mentioning the reemergence of the Arabs and Islām and the Middle East as a significant world force, the rise and fall of the Soviet Union, and the immense and profound impact of science and technology on the whole of mankind.

This is no book for the scholar. I expect no scholar worth his salt could pass my selections without making a very good case for different ones. On the other hand, I believe the selections here would have to be included in any definitive future "History." The material is severely exclusive by design. There is no intent to present a complete 20th-century history, even within the selected subject areas. Nor have I intended to present a detailed, or even complete, pre-20th-century history. Some specialists and history buffs will be amazed at what I have left out. I too was amazed, and pained; it would have been much easier to make this book 800 pages instead of 300, but I believe that beyond a certain point, length would have obscured the key vision, the tidal sweep of events. My selection criteria have been relevance to current events that illustrate this tidal sweep, and narrative coherence. Beyond these I have merely striven for factual accuracy and interpretive integrity with occasional garnishes to preserve the natural interest of the subject. In the spirit of brevity and focus I've dispensed with the traditional flood of footnotes and references since their distraction is contrary to my intent. Nevertheless, it would be a shame to present this story without at least a few leads into the immensely richer detail beneath. Accordingly, listed under "For More Information" are some works that I've found to be particularly interesting or useful. Alas, some of these are no longer in print but are frequently available in used bookshops or in the library. Beyond all this it is important to understand that my simplification necessarily fails to represent the intricacies of real life and history. We are interested here in the tides of history, not the complex froth and splashing waves of its parts.

A few words are in order about usage and terminology. When I write of "Islām" or of "Muslims" I will usually be referring to the relevant world and peoples of the Arabian Peninsula, the eastern Mediterranean and eastward to Pakistan. Unless explicitly indicated otherwise I will be excluding the peoples of Africa's interior and north coast west of Egypt. I have left out these and other Muslim countries, indeed more than half of the world's one billion Muslims, because the tidal dynamics aren't there, and because an important part of that dynamic, fundamentalist Islām, is largely a religion and culture of the passionate Arabs and Persians.

In Part II where I write of the Russian "nobility" I have in fact gathered together several privileged, overlapping groups that specialists would define separately as the gentry, the aristocracy, the minor nobility, and the nobility. Also, while the Russians have always traditionally referred to the "Tatars" or "Tartars" I often use the term "Mongols" because of its more universal use. It seems to me this is appropriate given the more universal impact of these invading nomads. A similar rational is behind my use of the term "Vikings" instead of the Russian usage, "Varangians."

Part III will differ somewhat from Parts I & II of this narrative because of the relatively unfamiliar quality of the subject material, after all, few of us can be expected to have much acquaintance with cosmology, electronics, nuclear science, fractals and chaos theory, or the space enterprise. On the other hand, I fear that without some description of the content of these areas it will be virtually meaningless to talk of their history. Accordingly, as we follow the flow of scientific discovery and technological development in Part III we will also touch on some of the relevant principles along the way, hopefully in a manner pleasantly accessible to the reader.

In telling the story of the revolutions of science and technology I have chosen five areas, not only because they are broadly representative of the scientific enterprise but because they are among the most historically significant; cosmology for its impact on our view of our place in creation, electronics because of its communications, control, and information facility, and total permeation of our lives, nuclear science because of the power it has put in our hands and the deep questions this power has forced us to face, the science of fractals and chaos which in barely a decade has begun to redirect the very conceptual foundations of the scientific community, and space for its obvious historical transcendence. Chemistry, molecular biology, medicine, elementary particle theory could easily have been included, but I don't think they have yet quite the significance of the first five. Clearly, however, one could argue a strong case otherwise.

For the first half century of nuclear science I have usually used the word "atomic," as in "atomic bomb" and "atomic energy" etc. where "nuclear" would be more accurate, because that was the common usage at the time. For events after 1950 I've used "nuclear" since by then this more accurate term was gaining currency.

Now, let's get on with it, and take a dip in the tides of history.

PART I

The Reemergence of the Islamic Middle East In World Affairs

For several hundred years the Muslim world has played no significant part in shaping global history. This is particularly striking in the Middle East. The final gasp of the Ottoman Empire, that last pretension of politically organized universal Islām, began about a century ago as it withdrew from the Balkans, and ended with the disaster of the First World War. Thereafter, the Islāmic world has survived as a fragmented curiosity. In the centuries immediately before its final collapse, Muslim lands were of interest mostly to Western explorers and other romantics seeking the world's backwaters for thrills, adventure, and occasional profit and easy political power in empire.

This was not always the case. At a time when Europe was as backward and barbarian as any place on earth, there was, conversely, a civilization as dazzling as any on earth, the civilization of Hārūn al-Rashīd's Abbāsid Caliphate in Baghdad, mostly known in the West through the romantic tales of *The Arabian Nights*. And just as the Romans looked upon the Vandals as utter barbarians sacking their capital, so too did the Muslims, with much better reason, view the Crusaders attacking Jerusalem. And, whereas the Vandals were virtually fatal to Rome, the occupation of Jerusalem was a hardly noticeable irritant to the Muslim world.

Nevertheless, the inevitable cycle of history took its turn, and the Muslim world splintered, battled itself, battled the world around it, and decayed until it became a curiosity and the European adventurer's toy.

Now the ebb tide appears to be over with the momentum of history just perceptibly beginning its deep and profound push in the other direction. The two World Wars and especially the world's current dependence on the oil resources of the Middle East are important factors contributing to the growing self awareness and political regeneration of Muslims from Morocco in the west to Pakistan in the east, and Syria, Iraq, and the Muslim states of the old Soviet Union in the north to Yemen and the Sudan in the south.

What is the source of this tide? And how powerfully will it peak? These are the issues we are now to encounter.

Chapter 1

Western Civilization
and
The Shadows of History at Its Fringes

In the *Muqaddimah*, written in 1377, the Muslim historian Ibn Khaldūn reveals a pattern of history: comfort is at once the pinnacle and death knell of a nation or civilization. Almost precisely 400 years later Edward Gibbon observed in his celebrated, *The Decline and Fall of the Roman Empire* that, "If a man were called to fix the period in the history of the world during which the condition of the human race was most happy and prosperous, he would, without hesitation, name that which elapsed from the death of Domitian to the accession of Commodus." Together, Ibn Khaldūn and Edward Gibbon are telling us that with the death of the philosopher emperor, Marcus Aurelius, in 180 A.D. the onset of the European Dark Age was coming into view, and that major changes in civilization could be expected.

Barely a century later the Emperor Constantine converted to Christianity thrusting aside not only Roman polytheism, but the Roman capital as well, transferring the Empire's political and geographical focus from Rome to Byzantium 900 miles east on the forested hills overlooking the Bosporus. By 395 the Roman Empire had permanently fractured into east and west. The west lingered in confusion another 81 years during which time Rome was captured by the Visigoths in 410 and sacked by the Vandals in 455. After 476 all that remained of the Western Roman Empire was a memory and the religious influence of Christianity.

The surviving Eastern Roman Empire found its political power had also evaporated. The distant provinces of Brittania, Hispania,

Gallia and Mauretania, as well as most of the rest of Europe and the coast of North Africa were no longer hers. What finally remained was what today constitutes Greece, Egypt, Turkey, Israel, Lebanon, Syria, and the western part of Iraq. By the early seventh century the Eastern Roman Empire had long since ceased to be either an Empire or Roman. Known by then as the Byzantine Empire it had its own parochial interests including commerce with the Orient and the protection of its borders further east where the Sasanids of Persia (Map 1) were constantly contesting its claims and slowly draining its resources in a never ending stream of little wars.

The early seventh century was a transition time for Sasanid Persia that put it in the same vulnerable position as the Byzantine Empire. Heir to a tradition over 2,000 years old, it's kings and wars are still read about in the Greek classics, *The Persian Wars* by Herodotus and *The Peloponnesian War* by Thucydides. By 620 the Sasanian dynasty had achieved considerable military and political success, but the next dozen years of war with Byzantium, and internal political chaos, finally exhausted it. Persian territories extended from modern Afghanistan in the east through the Iranian Plateau, down upon the fertile planes of Mesopotamia and across the Tigris and the Euphrates rivers right up to the border of the Byzantine Empire. North and northeast lay the gut of Eurasia, always ready to debouch its seemingly endless supply of barbarians upon the civilized world, and to the south, the arid frontier disappearing into the deserts of Arabia out of which troubling but manageable Bedouin raiders periodically pestered.

In a quirk of history it was these insignificant Bedouin tribes bursting out of the desert, unified and driven by religion and the promise of easy plunder, that changed the course of history more profoundly than any Hun, Vandal, or Gothic hoard from the north.

Arabia! Did ever any name conjure an image more remote, of empty and mysterious dominion, of trackless, sandy, stony wilds than Arabia. Arabia stood then as it stands now, a rectangular plateau, 700 miles by 1,200 miles, rising abruptly in the west along the Red Sea and sloping eastward to the Persian Gulf with a narrow fringe of fertile land here and there along the coasts and in the nearby mountains. Encompassing an area as large as the United States west of Denver with a Spain-size spur on its south

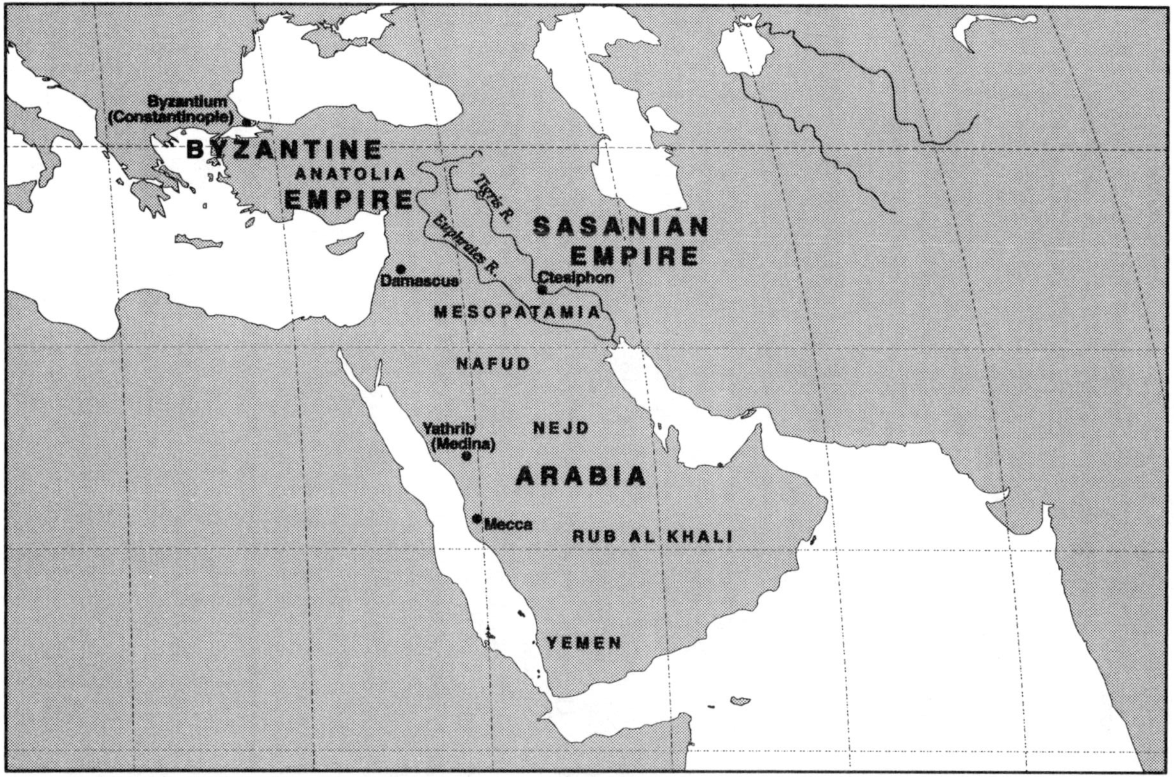

Map 1
The Middle East of Muhammad's Time

east corner, it has not a single permanent river or stream. The dotted blue water courses shown on maps merely show the *wādīs*,epressions along which water flows when it rains. The annual rainfall is comparable to one or two good summer thunderstorms in Iowa. Temperatures range beyond 120 degrees in the summer sun to below freezing on winter nights. Like the plains of South Dakota the wind blows constantly, and the Bedouin take what little advantage they can of its dryness and evaporative cooling qualities. In the central Nejd plateau and the great deserts of the An Nafūd to the north and the Rub' al Khālī to the south sand and rock dominate, although even here the careful observer or hungry animal can find pasturage. For man and animal accustomed to these climes life is hard and hardening.

Millennia of struggle had produced a lean and hearty people. Tough, generous, proud, hospitable, loyal, and fiercely tribal, blood truly was thicker than water, an important matter in a land so parched. Part of the tribal identity was also in its gods. Bedouin polytheism was as rich as tribal imagination, geography, and geology demanded with spirits identified with many trees, rocks, springs, stars, and places. The desert tribes known then as the Kalb, Ghatafān, Hawāzin, Aus, Hārith and others, as well as the Quraysh, each had a unique pantheon sharing little with their neighbors except perhaps a principal deity or two, among them one known as *al-ilah*. Intertribal conflict over water, pasturage, and honor, always a fact, was exacerbated further by religious loyalties.

Mecca was the one place in Arabia free of this constant tribal conflict, a place of pilgrimage where many tribal gods including *al-ilāh* were represented in the Ka'bah. The Ka'bah is the central shrine of Islām, but predates Islām by at least 700 years with the first known written mention of it about 60 B.C. by the Roman historian Diodorus Siculus. It is a stone building, roughly cubic, 40 feet long, 35 wide and 50 high with no windows and a single double leaf door seven feet above the ground. The door is accessible by a wheeled, steep wooden staircase brought over from the nearby mosque as needed. Outside, at the southeast corner, embedded shoulder high in the wall is the sacred Black Stone, probably of meteoric origin, that has been revered by the Bedouin as their most holy object since before the birth of Christ.

And so it continues to this day with the world's Muslims.

Mecca, located 50 miles inland from the Red Sea and 650 miles south from the tip of the Sinai, was by the late sixth century not only the center of pilgrimage in Arabia but an important trading center as well. It was well located to take advantage of the instability in Persia and Byzantium where constant warfare had forced commerce away from the direct routes to the Orient to a circuitous path down the length of Arabia to Yemen, where goods were shipped by sea to and from India and China.

The Quraysh tribe, who about the year 500 had moved into the area, were very successful in their forceful control of the spice trade. Their technique reminds us of modern mobsters or corrupt politicians who sell 'protection' and influence. For a fee the Quraysh would see to it that the caravans passing through their area were free of bandit or tribal harassment. Other fees would get you the benefit of their contacts with merchants up north in Damascus and shippers down south in Yemen. The situation turned out so profitable that many of the Quraysh moved to town and settled in Mecca, where naturally enough, they were also in control.

About this time a feud broke out in the tribe between two otherwise anonymous figures, Hāshim and his nephew, Umayyah, that has had ramifications felt even into the 20th century. The followers of each split into the separate clans Hāshim and Umayyah. The Umayyah clan became dominant in Mecca, and eventually one of its members founded the Umayyad Caliphate, the first ruling dynasty in Islāmic history. The Banū Hāshim lost influence and became rather second class citizens of Mecca as the animosities between clans continued. Nevertheless, the Banū Hāshim have had at least two illustrious sons, the most recent, King Hussain of the Hāshemite Kingdom of Jordan. The other was Hāshim's great grandson, Abū Al-Qāsim Muhammad Ibn 'Abd Allāh Ibn 'Abd Al-Muttalib Ibn Hāshim, known to history simply as Muhammad.

Who could have guessed that the child Muhammad, fatherless at birth and orphaned by six, member of an inferior if not oppressed clan in a backwater town deep in the wastes of a land barely on the fringe of civilization, would become among the two or three most significant persons in human history?

Born in Mecca about 570, he came to an Arabia seething with unharnessed energy. The Bedouin tribes were raiding each other and the empires to the north at the same time they were facilitating the caravans between Damascus and Yemen, and between Mecca and the Persian Gulf. The polytheistic paganism of the tribes boiled in a stew laced with many messianic prophets predicting the coming of one who would destroy the tribal pantheons and replace them with the one true god. The *hanīfs*, neither Christian nor Jew, were preachers of monotheism that was beginning to weaken tribal polytheism. Jews and Christians with their monotheistic religions were also no strangers to Mecca or to Yathrib (Medīna), the other principal town in the region.

Some evidence suggests the young Muhammad was exposed to Jewish and Christian beliefs while accompanying his uncle on caravan to Syria, but the potential for local exposure was entirely adequate to account for his later knowledge of the Talmud and Bible however fragmented and imperfect that knowledge may have been. Waraqah ibn Maufal, a cousin of Muhammad's first wife, Khadīja, was a well known *hanīf* who must have spoken often with Muhammad, and probably influenced him. Nevertheless, for the first 40 years of his life what Muhammad saw and heard was simply absorbed, provoking no action, only thought.

He appears to have had a contemplative habit, for he was known to frequently retire to a cave in the hills outside of Mecca where he would go to think. Judging from later traditions and the Qur'ān he must have been troubled by the constant intertribal blood feuds as well as the religious conflicts driven by tribal loyalties to unique gods. Some have also suggested that his later revelations may have been influenced by a growing disregard of the poor by the increasingly prosperous mercantile class of Mecca. Muhammad himself was in a good position to appreciate both worlds, coming as he did from a modest background and eventually progressing, through talent and luck, to a comfortable living.

"Oh, Muhammad thou art the messenger of Allāh, and I am Gabriel." Thus Muhammad learned of his mission in the year 610. This first vision occurred while on retreat at his mountain cave. The words come down to us in the *hadīth* or traditions of the

Prophet's life collected and recorded by his followers. Just preceding this utterance Muhammad was commanded in a vision to read some Arabic writing on a silk cloth. He protested that he couldn't read. The archangel insisted thrice more, and lo, Muhammad read, "Read in the name of your Lord who created .." These are the first words collected and later written in the Qur'ān, a book whose influence in human history is comparable only to that of the Bible. (However, for reasons that are not clear these words appear in the Qur'ān very late in the book, in the 96th chapter, "The Surāh of the Clot.")

For three years Muhammad said nothing publicly about this vision or any of the others he subsequently had. But privately he did persuade his wife and a few friends and relatives to discard their traditional tribal beliefs, accept the message of Allāh, and submit to His will ("islām" means submission/surrender in Arabic.)

About 613 he began public preaching, slowly gathering converts in Mecca. At first he was ignored by the town leaders, then looked upon as a harmless eccentric. But as his message that there is "no god but Allāh" got wider public attention the mercantile interests began to feel threatened. It simply would not do, religiously or commercially, to have the sacred deities of the Ka'bah thrown out. Business would be ruined. (Indeed, some modern scholars have conjectured that had not Mecca been the religious center that it was it might have gone the way of Petra, that amazing ghost city carved out of stone in a rocky basin and surrounding gorges in southwestern Jordan and largely abandoned one and a half thousand years ago.) At one point, in trying to accommodate the Meccans, Muhammad had a revelation telling him that it was alright to continue worshipping three other deities, the very popular al-Lat, Manat, and al-Uzza, thought to be the 'daughters' of *al-ilāh*. However, he very quickly appreciated this inconsistency with Islām's central idea that "There is no god but Allāh" and recanted, telling his followers that he had been deceived by the Devil. Accordingly, these verses of the Qur'ān were expunged, and have ever since been known as the 'satanic' verses. Of modern relevance, Salman Rushdie's 1989 novel "Satanic Verses" deals in part with these dethroned goddesses. The extremely zealous

Shī'ites of Iran were so offended by this that they declared a death sentence on the author. The recantation almost had similar results for Muhammad for it precipitated a final rapid decline in his relations with Mecca's ruling class until he and his followers were forced to flee town. He arrived in Yathrib on September 24, 622. This emigration is called the *Hijra* (Hegira) and marks the year from which Muslims record all subsequent events in history. Accordingly, the date of Muhammad's death in 632 A.D. is recorded by Muslims as 11 A.H. (Anno Hegirae), or the year A.D. (May 1995 through April 1996) as 1416 A.H..[1]

Fortunately, by 622 he was sufficiently well known in the surrounding area that he was welcomed at Yathrib. Indeed, he was invited there by representatives of several Arab clans to mediate a local blood feud. Within a short time a seminal event occurred, driven by the necessities of the situation, which has, arguably, more characteristically imprinted Islāmic culture than any other thing. In resolving the blood feud, many had joined the emigrants from Mecca and converted to Islām. In order to maintain this group with any hope of long term harmony, Muhammad declared that henceforth all believers form a single community, the *Umma* (the community of believers) and that commitment to this community and its rules superseded all previous allegiances to family, tribe, or nation. The shift of allegiance from blood ties to Allāh was profound; this was at once a religious and a political commitment, a fusion of loyalties that was unique and quite contrary to that generally accepted in the West of Christ's distinction when He said, "Render therefore to Caesar the things that are Caesar's, and to God the things that are God's." This commitment, at least in principle, by Muslims to the unity of religion and politics, and indeed, to the unity of all of life's endeavors has had consequences for Muslims and non-Muslims of historic scope which we shall encounter again and again in our

[1] The Muslim calendar uses the lunar year of twelve lunar months equaling 354 days. Since the solar year on which the Gregorian calendar is based has about 365 1/4 days the Muslim calendar gains a year every 32 1/2 years.

story.

As his influence and following grew he came into conflict with the Jews of Yathrib who rejected his assertion of being the latest of a long line of biblical prophets, and alas, as so often happens in history, the zealous advocate of God's will resorted to the sword. The subsequent shock to the city's economy in destroying the centrally productive Jewish clans drove Muhammad to another decision; thereafter, he and his followers were to give captured unbelievers a choice of conversion, death, or tribute. Uncounted thousands of Christians and Jews chose to pay, and having paid, were left to go about their lives unmolested as continuing residents of the conquered lands. Through the next 12 centuries this was generally the policy of Islāmic States, a remarkable historic fact in view of our 20th-century experience otherwise.

The sword also served Muhammad beyond the environs of Medīna ("the city", for so Yathrib was becoming known) as his Muslim raiders began to attack caravans bound for Mecca 200 miles to the south. Naturally the merchants of Mecca were upset by this; in the first place they were losing money, and secondly, they were becoming keenly sensitive to Muhammad's growing power. Eight years of occasional battle between the cities, with armies sometimes amounting to more than 10,000, each side alternately winning and losing, finally decided the issue in favor of Muhammad and his followers. A treaty was signed and eventually a triumphant Muhammad returned to Mecca in 630.

There followed a nobly mild reprisal against his onetime enemies as well as the reassuringly good example of his Muslims treating tribal sensitivities at the Ka'bah with at least some tact and discretion. The Ka'bah was indeed purged clean of its pantheon of idols, save two, the sacred Black Stone and *al-ilāh,* known ever after as Allāh; "the god" had now become "God."

Word of Muhammad's military successes over the decade 622 to 632 became known to the tribes the length and breadth of Arabia. At the cost of much blood some learned of it first hand. Others heard the message and voluntarily choose to submit. Nothing succeeds like success, and in a land like medieval Arabia success at arms was especially successful. Not only did such success bring power, it also brought booty and the message and promise of Islām. When contrasted with the tribal religions Islām presented a unity, harmony, and simplicity totally beyond the vague offerings of polytheism. However, above all else stood Islām's prodigious power to politically unify the tribes. In the harsh, elemental world of the Bedouin the prospects of Islām, power, and booty must have seemed irresistible. When Muhammad died of natural causes in 632 in Medīna all of Arabia was Muslim, and the empires of Byzantium and Persia were on the threshold of disaster.

"There is no god but Allāh, and Muhammad is his Prophet." This declaration, sincerely made before a Muslim, is all that is required to become a Muslim. The practice of Islām is almost as simple, including what are known as the "five pillars"; 1./ the declaration of faith given above, called the *shahādah*, 2./ prayer five times daily, called the *salāt*, 3./ pilgrimage to Mecca at least once in one's lifetime, called the *hajj,* 4./ the fast of the month of Ramadan where neither food nor water are taken nor sexual intercourse indulged in from sun up to sun set, called the *sawm*, and 5./ alms of 2½ to 10 percent of one's income given to charity, called the *zakāt*. This is a small price to pay for everlasting pleasure in Paradise. But of course, there are a few other details determining a Muslim's behavior.

The Qur'ān is the supreme authority in Islām. It is the eternally existing book transmitted from God by the angel Gabriel to Muhammad in the Arabic language. Unlike the Bible, which was <u>inspired</u> by God, the Qur'ān <u>is directly</u> God's word (brought by Gabriel). <u>The distinction is the essence</u>; the "inspired" may be modified, for example through translation from original texts. The Qur'ānic word of God cannot be modified in any way because it would be tampering with Allāh Himself. That's one reason why the Arabic language was so universal and essential in Islām, and why for centuries translation of the Qur'ān was not encouraged. The Qur'ān originally existed in fragments, some written, some only in the memory of those who heard Muhammad's revelations. The fragments were first collected and committed to writing about a year after Muhammad died. The standard version we know today with its peculiar arrangement of chapters (surāhs), not chronologically but by length with the longest first, was put together about 20 years later.

The *Hadīth* is a record of the words and actions of the Prophet as remembered by those around him, a kind of composite biography. It is composed of several thousand episodes, each like a short reminiscence; for example, the story of Muhammad's first vision of the angel Gabriel mentioned earlier. The *Hadīth* is second only to the Qur'ān in authority among Muslims. In the centuries following Muhammad's death *hadīth* were frequently counterfeited so that by the ninth century the number of *hadīth* had grown to more than half a million items. Muslim scholars at that time made several independent, critical reviews of the *hadīthic* literature and settled on the *Hadīth* known and accepted by Sunnī Muslims today. It is the basis for establishing the *Sunna*, the orthodox customs and practices for the Muslim community. A somewhat different version is accepted by Shī'ite Muslims.

The *Sharī'a* is the legal system of Islām. It is the divinely supported, indistinguishably religious and secular Law of behavior. Although based on the Qur'ān and the *Hadīth*, it was man made and often developed in the absence of specific Qur'ānic or *Hadīthic* guidance. Nevertheless, it is seen as an expression of God's will concerning Muslim behavior. This is quite the opposite of Western law which is experience-based, pragmatic, and claims no divine authority. Saudi Arabia and Iran are examples of late 20th-century Islāmic States living under strict compliance to the *Sharī'a*. The *Sharī'a* illustrates the principle that in Islām there can be no distinction between church and state; the Qur'ān and the *Hadīth,* through the *Sharī'a* determine what is lawful and what is not for all actions. On the other hand most Islāmic States do not, and have not, strictly adhered to the *Sharī'a*.

There is no clergy in Islām the way there is in Christianity. Muslim leaders never have a sacerdotal role, even in the mosque. They never intervene between the believer and Allāh (here too, the Shī'ites believe somewhat differently with their Imām playing a special role). In the West the term "clergy" or "cleric" is sometimes applied to Muslim leaders, especially those aggressively practicing their faith and therefore involving themselves in all aspects of public life. There have been many figures like this that have had public influence and leadership functions, the historic nature of which can be understood by their titles, such as caliph, sultan, imām, and ayatollāh. The Caliph was the successor and deputy of Muhammad, the leader of the Muslim state with complete religious and secular powers. He led the prayers in the mosque as well as the army in the field. As we shall see later, the caliphate declined and died, gradually being replaced by the sultanate. The Sultan was mainly a military/political and rather secular ruler though at times some tried to assume religious leadership as well. The imām was a special kind of leader with the term having strong associations with divine favor or mission.

Muslims trace their traditions back through the Old Testament to Adam who is supposed to have received the Black Stone from God as a token of forgiveness when he was expelled from Paradise. Further tradition has it that the first Ka'bah was built by Abraham and Ishmael. Another link with the Old Testament is seen when Muhammad was in Medīna; he first enjoined his followers to turn to Jerusalem to pray, and only later, after his conflict with the Jews, did he change the direction to Mecca. With these examples and those mentioned earlier it is clear that, contrary to the belief of many Christians and Jews, Islām is a natural continuation of the Judeo-Christian tradition. It has certainly always been viewed as such by Muslims. (The Muslim position is strikingly reminiscent of a much later development within Christianity, the appearance of The Church of Jesus Christ of Latter-day Saints (Mormons) with its prophet and holy book, Joseph Smith and the Book of Mormon.) Muslims view Christians and Jews as, "People of the Book." They cherish and revere the Old Testament and its prophets just as Christians and Jews do. Indeed, they believe Muhammad is the most recent of the long line of prophets that includes Jesus. And therein lies Islām's chief quarrel with Christianity; Muslims are scandalized by the Christian deification of Jesus, and they are even more scandalized by the trinitization of God...... there is only <u>one</u> God .....ie., "There is no god but Allāh..." As for the Jews, historically the Muslim quarrel has been with Jewish refusal to acknowledge Muhammad's primacy among the prophets.

Muslims do not regard Muhammad as divine nor do they pray to him. Muhammad himself always insisted that he was just an ordinary mortal chosen to act as God's messenger, and he took great pains to discourage those who wanted to elevate him beyond this level. His immediate successor, the first Caliph, Abū Bakr, thoroughly understood this and articulated it to reduce the confusion and panic at Muhammad's death; "Whoever worships Muhammad, know that Muhammad is dead. But whoever of you worships God, know that God is alive and does not die."

Chapter 2

Muslim Expansion, the Golden Age of Islām, Its Beginning and End

"Muhammad is dead!" Imagine the confusion of these simple, seventh-century Arabs who had surrendered their lives and their very souls to the guidance of one man. His message and promise of Paradise were contingent only on this surrender, and though he had frequently warned that he was merely the messenger, his listeners could not help but revere him, the tangible manifestation of an abstract idea. Some did not believe that he was dead. Others were at a loss at what to do with their lives if he was gone.

Amongst Muhammad's inner circle a deeper and clearer vision of his role was apparent, "Muhammad is dead. But ... Allāh is alive and does not die." How then shall the community carry on? Muhammad had not even mentioned the matter of a successor. A gathering at Muhammad's house produced a quick consensus that his father-in-law Abū Bakr was the best man to succeed in Muhammad's role as leader of the Muslim community, especially as he was a Sunnī, a strong and respected follower of Muhammad's practices, the *sunna*. Only the day before he had been asked by The Prophet to lead the prayers of the day in his stead. This successor or deputy (*khalīfa*) we call the Caliph.

The consensus was not unanimous. Alī, cousin of Muhammad, who had also been adopted by Muhammad, and later also married his daughter Fātima, bitterly resented being passed over. He and his followers, later known as Shī'ites (the party of Alī, *shī'at 'Alī*) felt that the succession should stay in the family. Once again the powerful Bedouin notion of blood ties was about to seed conflict, this time not only by shadowy tribes in the nameless deserts fringing recorded history, but at the very center of history.

And the fruits would reappear again and again and again for more than 1,300 years up to our own time; Sunnīs against Shī'ites.

Like all movements, Islām was very vulnerable in its early stages and might easily have fragmented and disintegrated without continued strong leadership. The vigorous move of the Muslim leadership in appointing Abu Bakr protected the movement and directed its energy. Like a lens gathering the sun's rays and focusing them to create a very hot spot, Arab energy was coming to a focus through the unifying power of Islām. The caliphate was the lens for this energy, the natural consequence of the need to have it controlled and directed.

Alī put aside his disappointment and joined the Arab armies in their conquests. In a short time given the comparatively small size of their army, they attacked, subdued, and pillaged Sasanid Persia, and much of the Byzantine Empire. Palestine and Syria were conquered by 640, Persia and Egypt a year later. Eastward expansion continued to the edge of China crossing the Oxus river in 667, through fabled Bukhara and Samarkand to the Jaxartes river in 712, and down into India in 713. It appeared nothing could stop the sword of Islām. Further conquests continued in the west along the coast of North Africa all the way to the Atlantic and across the Straights of Gibraltar up into Spain, parts of which would remain under Muslim domination for more than 700 years. Not until 732, a hundred years after Muhammad's death, were the Muslims finally stopped with their defeat by Charles Martel, Charlemagne's grandfather, at Poitiers, less than 200 miles from Paris. It may have been the depth of the Dark Ages for Europe, but it was a heady time for the Arabs and the Middle East (Map 2).

While all this was happening at the expanding edges of the Muslim Empire there was almost as much action at the center. The first four Caliphs were variously known as the "Rightly Guided" or the "Companion" Caliphs, because they had known the Prophet personally. Abu Bakr was elected in 632, and died two years later of a fever. Umar lasted 10 years before being assassinated by a Christian slave. Uthmān lasted 12 years before being assassinated by a group of unhappy Muslim soldiers. Alī finally became Caliph

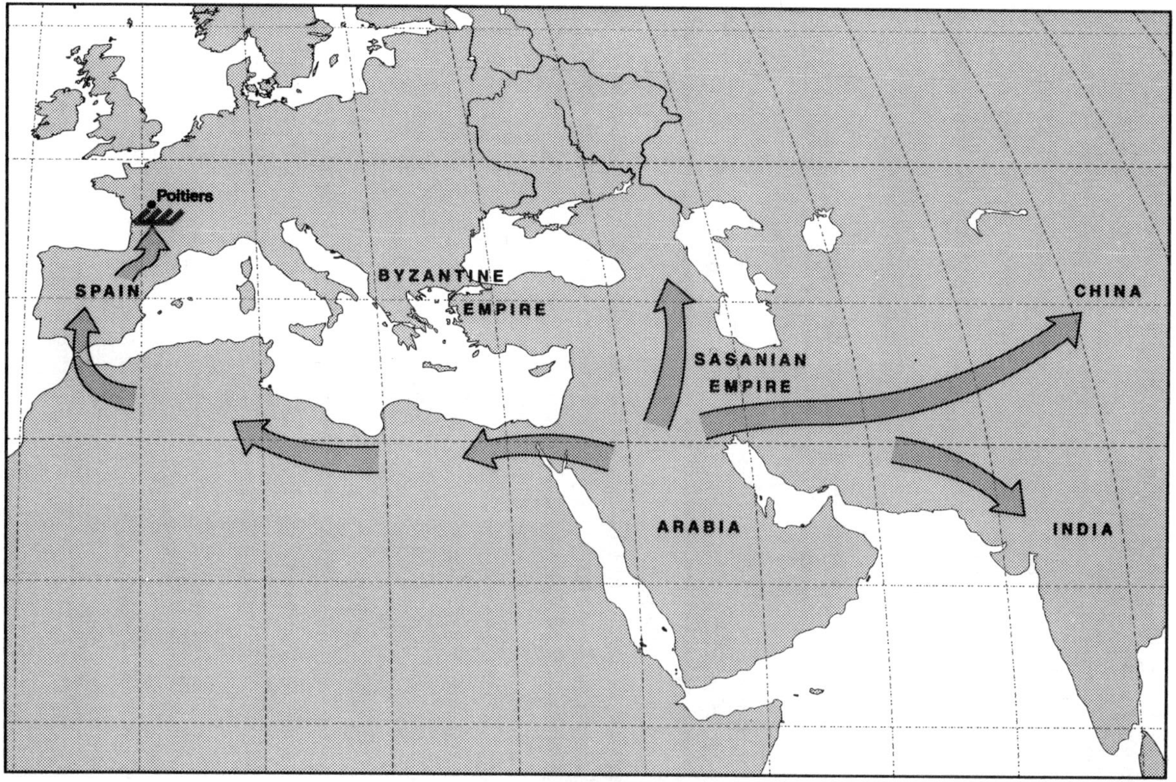

Map 2
Muslim Conquests

in 656, but survived only three years before being assassinated by another unhappy Muslim. In the first 27 years after the Prophet's death three of the four Caliphs were murdered. What is particularly revealing is that two of them were murdered by fellow Muslims, yet the Prophet had specifically forbidden Muslims from killing Muslims. Muhammad's moral prohibitions have been no more a deterrent to Muslims than "Thou shalt not kill" has been to Christians and Jews. Like the moral power of Christianity and Judaism, the moral power of Islām has always been notably limited in the face of politics or passion.

However, not only Caliphs were being slaughtered by fellow Muslims. In 656 Alī and his forces fought the "Battle of the Camel," against other Muslims in order to subdue those that didn't want him as Caliph. (It was called the "Battle of the Camel" because the influential Ā'isha, one of Muhammad's later wives, urged Alī's defeat while watching the fight from a litter atop her camel.) For the remainder of his caliphate the story is the same, battles between Muslims over who will lead the Empire. Coincidentally or not, many of Alī's adversaries were of the clan of Umayyah who 150 years before had feuded with Alī's clan Hāshim. By 659 the Umayyads, under the leadership of Mu'āwiyah, then governor of Syria, not only controlled Mecca but Syria and Egypt. Mu'āwiyah was a nephew of Uthmān and claimed the caliphate over Alī. Alī's murder probably saved the Islāmic world from considerably more bloodshed in the short run, but in the long run saved nothing, for Sunnīs and Shī'ites would continue the bloodletting even to the Iran/Iraq War of the 1980s and the persecution of the Iraqi Shī'ites by Saddam Hussein in the early 1990s.

Alī's death permanently ended the possibility of a Shī'ite reign over the Muslim world. Mu'āwiyah founded the Umayyad Caliphate, and moved the administrative capital of Islām north out of Arabia to Damascus. It would never return. The first unconscious step had been taken removing the Arabs and Arabia from control of the Muslim world. Simultaneously, the followers of Alī and his sons declared the Caliph in Damascus unlawful, thus creating the first major split in Islām as they shifted their religious allegiance to Alī's descendants, the Imāms.

The Imāmate is one of the features distinguishing Shī'ite and Sunnī Islām. For Shī'ites the leader of Islām, the Imām, must be of Muhammad's family, and was, accordingly, divinely appointed. He is also viewed as sinless, the possessor of special, divine knowledge, and the sole and infallible interpreter of doctrine.

For Sunnīs, the term "Imām" was synonymous with "Caliph" (it was first used in the Qu'rān as "pattern," to be followed as an example; a leader) and later often used as an honorific title or for any leader of the Muslim community. He was the prayer leader in the mosque. The Sunnī Imām was appointed by men, not God, and was accepted as a sinner and liable to error like the rest of us. Community consensus, not the Imām, determined and interpreted doctrine. Nevertheless, he was an authority figure and was expected to be obeyed.

Accordingly, while the Caliph reigned in Damascus and elsewhere for the majority of Muslims, the Imām ruled for the Shī'ite minority, and just as there existed the institution of the Caliphate, there simultaneously existed the institution of the Imāmate for the Shī'ites. The most fertile ground for Shīism was at various times in Yemen, North Africa and Egypt, as well as in Iraq, and in Persia where it eventually became, and remains, the belief of the overwhelming majority. It's dominance in Egypt during the 10th through 12th centuries was manifest as the Fātimid Caliphate.

The conflict between the Caliph and the Imām was not merely a religious affair, it was also political and social. Since religion and politics were indistinguishable in Islām it has never been totally clear which drives the social engine on a given issue, religion or politics. Islāmic states are in theory a theocracy in which God is the sole source of power and law, so it was natural that political factions would take the form of religious sects, which have ever sense acted like our political parties.

The Umayyad Caliphate survived less than 100 years. During that time Islām's political boundaries expanded to the Atlantic in the west and to the border of China in the east; it probed India and Europe and established itself in Spain, not to be ejected till 1492. The Muslim state of Pakistan exists today as a direct result of the Umayyad conquests 1,200 years ago.

The Sūfī movement emerged during the Umayyad Caliphate, and thereafter fully permeated the Muslim world. To varying degrees the Sūfīs were persecuted, ignored, or courted, depending on the fears and needs of the ruler at hand. Paradoxically, the movement on occasion played an important role in the survival of Islām, for Sūfī brotherhoods provided social coherence when central government weakened. The Sūfīs (from the Arabic, *sūf,* for the wool of the simple garments worn in the beginning) originated with ascetics rebelling against the materialism and mechanical observances into which the Muslim faith rapidly fell as military and political successes mounted. Sūfīs preached a loved and loving God directly accessible through mystical experience. This was in sharp contrast to orthodoxy's view of a coldly formal God demanding the worshiper's submission, and in the case of the Shī'ites, to be approached only indirectly through the Imām. Sūfī practices eventually evolved from asceticism into mysticism, and from individual practice into membership in "orders" with study under a master. The Sūfīs are of interest to us because they were among the most influential in making converts to Islām, principally because they were very flexible in what they demanded often allowing the converts to retain substantial portions of their preexisting faith. This led to a wide spread folk practice of Islām far removed from the practices and beliefs of the *Ulamā*, the religious leadership, and in the process put considerable psychological distance between the two, palpably weakening the unity of the *umma*, the Muslim community. This is one reason why Indonesia the most populous Muslim nation in the world is so little involved with the Muslim travails in the Middle East today. In the past few hundred years Sufism has lost much of its energy, and it's not clear what role it will have in the decades to come.

The age of the sword, expansion, and Arab influence was about to end, and the most glorious age of Muslim culture about to begin. The link, alas, was more blood.

Why is slaughter so constant in this narrative? It's not an artifact of selection, for no historian could write this history without acknowledging it. Nor is it an aberration of the peoples of the Middle East, for Europe, the Far East, and the Americas have similar histories. Nor is it merely an artifact of the time, for men have killed men with appalling constancy from time immemorial. And what about the paradox we are about to reveal where the most wonderful products of man and his culture will spring from this blood saturated ground, a phenomenon not unique to the time and place of this narrative either. Are blood, conflict, and suffering prerequisites for growth and development? Are pain and suffering inextricably linked to fulfillment and happiness? This is not the place to treat these issues, but we might mention in passing Arnold Toynbee's conjecture that civilizations grow, fail, or succeed in accordance with their ability to respond to surrounding challenges. The dynamics of optimized challenge and response produce the most successful civilizations. Ibn Khaldūn 600 years earlier contributed the idea that cultural success springs out of conflict, and having succeeded, becomes weak and unable to resist subsequent competition. In the 20th century the Nobel Prize winning physicist, Max Plank, observed that, "A new truth does not triumph by convincing its opponents and making them see the light, but rather because its opponents eventually die, and a new generation grows up that is familiar with it." Though Plank was referring to the genteel world of science, the essence of his idea may be applicable to history in general; there may be limits to evolutionary social change beyond which only revolution can be effective. So let us not judge too harshly the Muslim engine of history for there is ample evidence that it runs on the same energy that drives us all.

The Umayyads were conquerors, not rulers, and it showed. Converts to Islām in the conquered territories, called *mawālī*, were in principle the equal of any Arab Muslim in the eyes of the Law. But in fact Arab racial arrogance seldom allowed real equality. Even the word, "*mawālī*," reflected this, for it means helper, freed slave, or client. The implied inferior rank is clear. Since many of the *mawālī* were Persians of relatively high sophistication compared to their Arab 'masters', and since the Arabs treated them as second class citizens, it was inevitable that before long the *mawālī* would become exceedingly unhappy with their lot under the Umayyads. Add to this the embittered Shī'ites and the still strong independent tribal instinct of the Bedouin to disregard central authority, even of the Caliph, and you get a situation too unstable to last.

In the late 740s resistance to the Umayyads grew and coalesced around Abū al-Abbās and his family who quickly dispatched the last Umayyad Caliph and established the Abbāsid Caliphate in 750. In a fury of slaughter he proceeded to kill all who might be pretender to an Umayyad claim to the Caliphate, missing only one, Abd al-Rahmān, who ran as far as he could, finally settling in Spain 3,000 land miles to the west and founding the Spanish Umayyad Caliphate in 756. Meanwhile, al-Abbās was so satisfied with his handiwork that he added the epithet *al-Saffāh* to his name. It means "the Bloodletter." Four years later he was dead of natural causes and succeeded by his half brother, al-Mansūr. Al-Mansūr further consolidated the family grip on the Caliphate, by the usual means, but also felt free enough to truly begin the process of Muslim refinement that would reach its zenith under his grandson, Hārūn al-Rashīd (786 - 809). Among the more significant things al-Mansūr did was to remove the Caliphate from Damascus in Syria to Madīnat as-Salām (City of Peace) on the west bank of the Tigris River in Iraq, which he began building in 762. It is interesting to note that in spite of the pretentious effort of al-Mansūr in naming the city the rest of humanity insisted on calling it after the nearby ancient village, Baghdad.

The Abbāsid destruction of the Umayyads marked not only a change in dynasty but a fundamental change in focus for Islām. The Umayyads were primarily Arab with a tribal outlook. Their capital was in Damascus with a corresponding interest westward to Byzantium and the Mediterranean. The Abbāsid capital was in Baghdad where the Persian influence with all of its customs and world view to the east was just short of overwhelming. Islām became universal, and before long left its Bedouin originators to return to their backwater in the Arabian peninsula.

We have remarked earlier about historic tides. Had a Muslim official followed the command of his Caliph and moved to the newly completed city of Madīnat as-Salām in 766, what might he have observed of the tide that was sweeping him along? Perhaps the first thing he would have noticed, even as he was approaching the city, was its design. As an official in the entourage of the Caliph he might easily have been a relatively well traveled Persian. The towns of Arabia and elsewhere in the frontier regions he may have visited were laid out with winding main streets from which branched many blind ways leading to the center of 'blocks' or quarters. They reflected the Bedouin heritage of free spirited disregard of urban life and structure, and never showed evidence of the rational urban planning that more sophisticated cultures developed. Yet Baghdad (Madīnat as-Salām) was different. As a Persian he would have found the wagon wheel like circular design with its four straight avenues radiating from the center very familiar, and in no way like previous Arab cities. Indeed, the city was designed by a Persian.

He would also have found a marked decline in Arab influence, beginning of course with the very location of the city only a few miles from the old Persian capital of Ctesiphon. Many of the administrators, like himself, were Persian. This trend, begun during the Umayyad Caliphate, became almost the rule in the Abbāsid Caliphate. The official demands of Empire simply exceeded an Arab's taste and lifestyle. And though Arabic remained the language of the Empire, the culture was quickly taking on Persian refinements. The Caliph was no longer a man of the desert, easily accessible, with simple tastes and simple pleasures. The Caliph and his court began to live and rule with a Persian splendor that would dazzle the most sophisticated courtier on earth. With each passing decade he became more remote until, by the ninth century, only the highest officers of state had access to him. The Grand Vizier, always it seemed, a Persian, became more than his deputy; he became the effective head of state, sometimes ruling the Caliph himself. This of course was not without its dangers to the Vizier; Hārūn al-Rashīd had his executed for becoming too powerful.

Our official would also have found that the Caliphs very quickly rejected a major element that brought them to power; there was to be no reestablishment of Shī'ite power. Shī'ite unrest had helped overthrow the Umayyads, but it was Sunnī orthodoxy that the rulers valued. In the second decade of the ninth century, al-Ma'mūn, son of Hārūn al-Rashīd, did try to end the Sunnī/Shī'ite schism by designating as his successor the 6th Imām (Shī'ite) instead of his own son, but extremist elements were still not satisfied, and the traditional green flag of Alī's descendants flew over Baghdad only a short time before the return of the black flag of al-Abbās.

Baghdad! Who can hear that name without a flood of images of oriental splendor, the romance of the *Arabian Nights*, and a city of very high culture and sophistication. Ninth-century Baghdad was briefly not only the most splendid capital in the world but the cultural and trade center as well. Its highly literate aristocracy accounted for many public and private libraries. Some individual ninth-century, private libraries held more books than existed at the time in all of Europe combined. We hear, for example, of the Vizier, Ibn Abbād, declining an invitation to visit another capital because of the excessive effort of loading the 400 camels required to transport his library.

With links to China, India, Persia, and ancient Greece through Syria, Palestine, and Egypt, the cultural resources of Islām focusing on Baghdad were breathtaking. Al-Mamūn (813 - 833) created the "House of Wisdom"(or "Institute of Science") and commanded its staff to translate into Arabic ancient texts gathered from across the empire and beyond its borders. He even sent requests for texts to Islām's perennial western enemy in Constantinople. Plato, Aristotle, Euclid, Archimedes, Hippocrates, Ptolemy, and Galen were only a few of the authors of antiquity that eventually became available in Arabic.

The West is deeply indebted to Islām as one of the most important intellectual bridges linking classical Greek culture to the European Renaissance. Many of the texts we have today are known only through their Arabic translations. However, the products of the maturing Muslim civilization were not limited to transmission of the Greek classics; other contributions included paper making brought from China, and the "Arabic" numbers, concepts of zero, and place value in counting by tens, all brought from India.

Over the next few centuries Muslims themselves made

important and lasting contributions to the world's culture. Architectural examples range from the still standing, seventh-century Dome of the Rock in Jerusalem, and the 13th-century Alhambra palace fortress in Granada to the 17th-century tomb memorial, the Taj Mahal in Agra, India. The architectural and calligraphic potential of the seductively flowing Arabic script has been sounded to amazing depths. To this day chemistry, anatomy, and optics retain elements developed during the Muslim flowering. In mathematics al-Khwārizmī's book on equations gave us the word "algebra" while al-Battānī extended and developed the trigonometric functions. These two are among the most important mathematicians of the Middle Ages and established algebra and trigonometry as we know them today. And in history modern historians stand in awe of Ibn Khaldūn's *Muqaddimah*. As geographers they measured the circumference of the earth at 20,000 miles, a figure considerably more accurate than the value known to Columbus several hundred years later, and described India, Ceylon, the East Indies, and Africa, as well as China more than 400 years before Marco Polo.

With imperial culture and comforts the Abbāsids began the all too familiar slide into decay and ruin. The two external sources of this ruin came in successive waves from the steppes of Asia. The first started as a trickle, individual slaves taken in the campaigns against the restless Turkomen tribes up on the northeast frontier. These slaves were excellent soldiers, and many eventually became palace guards. With the passing generations the Abbāsids became accustomed to them to the degree that many became officers. About 970 a more coherent tide began to move out of Turkistan and into the Muslim lands of Transoxania where they soon accepted Sunnī Islām. By the middle of the 11th century the internal Turkish influence as well as the successful invasion out of Turkistan rendered the Abbāsid Caliphate politically powerless. The Seljuk Turks asserted political and military power under a leader called the Sultan. However, the Seljuk Sultanate retained the Caliphs as convenient instruments to control the population through religion.

The influence of the Seljuk Sultan extended from Transoxania in the east to Syria, Palestine, and eastern Anatolia in the west. The Umayyads and their Moorish successors in Spain continued to ignore Baghdad and indeed had proclaimed their own Caliph. The political fragmentation of Islām scarcely stopped there however, for in Egypt and other parts of coastal North Africa the Shī'ite, Fātimid Caliphate was established which would stand for almost 300 years before being destroyed by Saladin in 1171.

The Fātimids originated directly out of the Shī'ite schismatic Sevener Imamate. With the death of the forth Caliph, Alī, Shī'ite leadership rejected the authority of the Sunnī Caliphs in favor of Alī's descendants, the Imāms. For the Shī'ite minority the Imām was far more than the Caliph, the Imām was the divinely inspired, infallible interpreter of Islāmic Law. Moreover, unlike the custom of the Sunnī majority where every Muslim could look directly to the Qur'ān for guidance, the Imām stood between the believer and his God. Because of his divine appointment he alone determined religious truth, and by Islāmic extension, secular truth as well. For a Shī'ite, the Qur'ān was no longer the direct guide for behavior, it was the Imām's interpretation of it. With such a fundamentally different view, conflict with the Sunnī majority was inevitable.

Normally the Shī'ite community coexisted with and within the Sunnī community, but they recognized the Imām, not the Caliph, as their leader. In most cases survival required that they avoid overt expression of this altered allegiance, and on occasion they became almost a secret society. Accordingly, for the Shī'ite, there existed a kind of shadow government awaiting its chance to overthrow the orthodox rulers. For a moment when the Abbāsids were coming to power it looked like their time had come, but the needs of political stability prevailed, and the Shī'ites were either suppressed or ignored as they returned to their practice of *taqīyah*, dissimulation.

The Shī'ite Imamate led by descendants of Alī survived for over 200 years ending with the death of the Twelfth Imām in 874. Its most significant schism occurred with the death of Ismā'īl, the Seventh Imām, in 760. Some refused to believe that he had died. They insisted that he had gone into hiding to become a "Hidden Imām" who would return as "Mahdī" to bring the message of salvation to true believers just before the end of the world. With the passing of time the Ismā'īlīs became the most radical element of Shī'ism spawning not only the Fātimid Caliphate but the cult of Ismā'īlī Assassins.

With such a pedigree the Fātimids behaved just as one might expect with unusual zeal in attacking their Sunnī brethren and exporting their brand of Islām. (The scenario, inspiration, and methods are familiar in our time beginning with the 1979 Islāmic Revolution in Iran.) For a time they controlled not only Egypt, where they founded al-Kahira (Cairo) in 969, but North Africa, Syria, Palestine, Yemen, and the Red Sea coasts including the holy cities of Mecca and Medīna. But all things end; by the end of the 11th century they had been fatally weakened by their continuous conflict with other Muslims, particularly the Seljuk Turks who were expanding at the expense of the Byzantine Empire and precipitating the Crusades. Then with a few jabs from the Crusaders the Shī'ite challenge to orthodox Sunnī hegemony over Islām ended with Saladin's destruction of the Fātimid Caliphate in 1171.

Neither Shī'ism nor Ismā'īlism was dead, however. The Persian Shī'ite community remained viable, if not prominent, in the continuing Imāmate. When the Twelfth Imām died in 874 the era of Imāms descended from Alī ended, and thereafter the Imām would be represented from generation to generation by his agents until his reappearance. The Twelfth Imām, like the Seventh Imām, was believed not to be dead, but "hidden" until his reappearance as Mahdī. It is from this Twelver "Hidden Imām" that the present leaders in Iran derive their spiritual and temporal authority 1,100 years later (Figure 1).

In an almost bewildering array of successive fractures reminiscent of Christianity's development, Islām split between Sunnīs and Shī'ites; Shī'ites split into Seveners and Twelvers; Twelvers split into the majority Shi'ites we now see in Iran and Iraq, and the Bahā'ī; Seveners split into Druze (whom we've seen fighting in Lebanon in the 1980s), Musta'līs, and Ismā'īlī Nazārīs.

The fracture of the Seveners into the Ismā'īlīs leads us to the

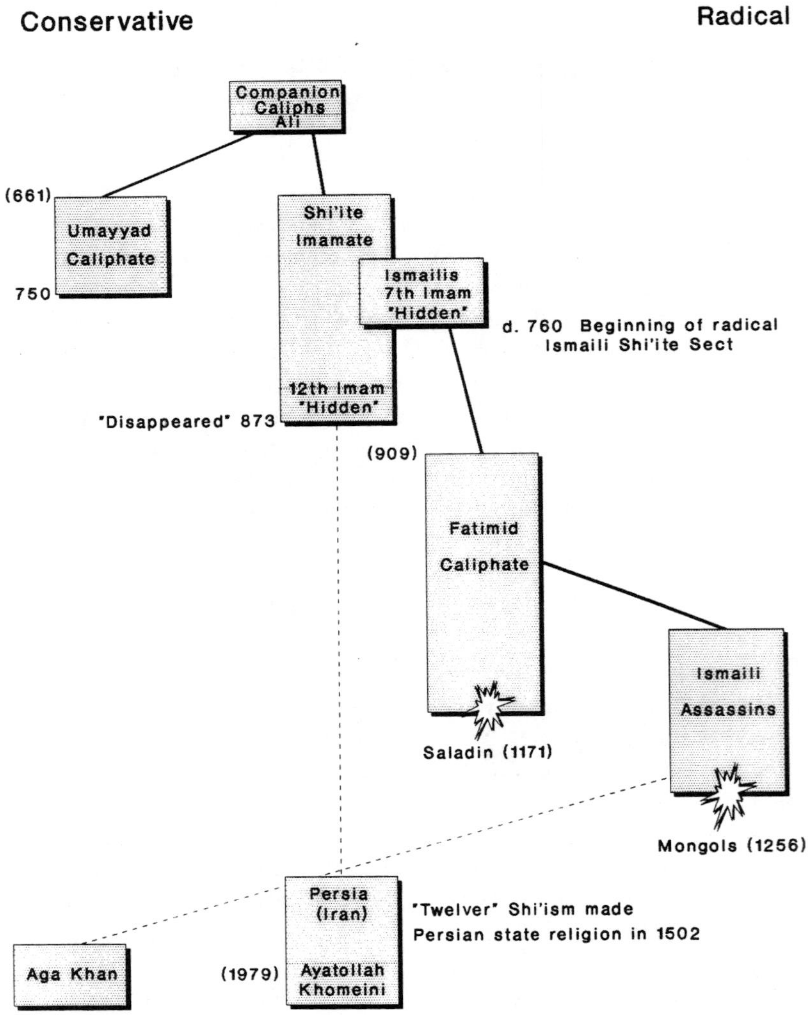

Figure 1
Shi'ite Family Tree

Assassins. In a dispute over the secession of a Fātimid Caliph in 1094, the Nizārīs faction took their own extremist path. They secured a number of fortresses in Syria and Persia where they exercised their influence throughout the Middle East by political intimidation and assassination. Some think they acquired the name "Assassins" because it was thought that their agents were recruited and sent on suicidal missions by being put in a compliant euphoria through the use of hashish, hence, hashish eaters, hence *hashshsāhīn*. After 200 years of political terrorism they made the fatal mistake of trying to extort the occupying Mongols in 1256. The Mongols invested and utterly destroyed their "impregnable" fortresses at Alamūt and elsewhere thus ending the centuries long reign of political terror in the Middle East.

In one of the fascinating turns of history the Ismā'īlī sect benignly survives, mostly in India and Pakistan, under the current leadership of the Aga Khan. Some today may be familiar with the image of the portly Aga Khan III sitting on one side of a scale while receiving a birthday gift from his followers of his weight in diamonds (which he promptly contributed to charity). One of his sons, the "playboy" Ali Khan, was better known in this country for his wife, the film actress of the 1940s and 1950s, Rita Hayworth.

Four hundred years after Muhammad's death the theoretical unity of Islām, the unity of life and the will of Allāh, as well as the unity of politics and religion, was in complete shambles. Indeed, almost all the major tenants of Islām were being ignored throughout the empire. Muslims were killing Muslims. Politics went its way ignoring religion, except where it was useful. Usury and idolatry were common. And wine, Qur'ānic prohibition notwithstanding, was not only enjoyed, but celebrated, as in the later, *Rubaiyat of Omar Khayyam*: "A Book of Verses underneath the Bough, a Jug of Wine, a Loaf of Bread - and Thou Beside me singing in the Wilderness - Oh, Wilderness were Paradise Enow!"

Where then was the coherence? Was there, or is there, a "Muslim World"? Yes, just like there was (is) a "Christian World." There was the common ideal manifest in the Qur'ān. There was a common script, and to a very important extent a common language. And notwithstanding the elastic treatment of religious principle, there remained many outward forms of religious practice, the daily anchor to a bedrock belief that helps individuals survive the random turbulence of history pounding in upon all sides.

Of special note, because of its current importance, has been the role of language to Arab/Muslim coherence. There are exceptionally few words from other languages in Arabic. This is all the more unexpected given Arabic's vast exposure to other cultures during the conquests and more recently to the modern dominance of western culture. There is clearly some resistance here, and its not difficult to see what and why. Arabic is the language in which Allāh, through His archangel, Gabriel, revealed the Qur'ān, and thus change has overtones of sacrilege. In addition, the Arab's oral literary tradition is their earliest and most cherished art form. The construction of new words from a rich storehouse of root words easily modified through the application of logical and systematic rules has rarely needed recourse to foreign terms for the expression of new ideas. The language is rich in rolling sonority, rhythms, assonances, and rhymes thoroughly appreciated by its speakers who value elegance of expression and brilliance of metaphor as well. For an Arab, language is not only an instrument of communication and an art form, it is a play thing, an occasion to enjoy the possibilities of wit. This pleasure in the use of the language has been manifest historically in his literature as well as currently in diplomacy. The Western style of communication sometimes misunderstands what is going on in the face of Arab loquacity, often taking it for stalling during international negotiations, as in the peace talks with Israel. In another recent instance the West understood Iraq's Saddam Hussein in the now famous "mother of all battles" statement as taunting and warning. That it may have been, but it was also, for the Arab, a natural exercise in the use of language to its fullest possibilities. On another occasion a couple of years later he said that he would, "Turn the skies into lava" if American aircraft tried to bomb Iraq. Putting aside the threat issue, one cannot help but marvel at the wonderful verbal image. These examples show that the Arabic language is a profound manifestation of Arab identity and as such has been protected from adulteration with the same motivation one occasionally sees in threatened interfaith or interracial marriages.

These examples also illustrate common and frequently misunderstood characteristics of Arab speakers as well. Exaggeration is part of the warp and woof of Arab communication. And exaggeration is often accompanied by emotionalism at the expense of the coldly analytical. Things are said that every Arab knows are not, and could not be true; it's just the way of talking. Saying something is often a substitute for doing it. On the other hand, repetition is a method of emphasis and a signal that the speaker is dealing with something important. In modern times these are important characteristics to remember as we listen to the dialog between the West and the Middle East.

We saw earlier the growing dominance of the Seljuks as an external source of Abbāsid ruin. The second external source of ruin was considerably more draconian than the Seljuk. Of all the great nomadic invasions in the world's history, none was more sweeping or more terrifying to their victims than the apocalyptic Mongols (Map 3). The prospect for survival of civilizations in the path of the inexorable Mongol catastrophe seemed as slim to the peoples of that era as we have viewed our chances for survival in a Superpower nuclear war in this era. And for millions in the 13th century it truly was the end of history. Urgench, the ancient capital of Khwārezmia, was literally wiped off the face of the earth when the Mongols enslaved or slaughtered every last inhabitant, burned the buildings to the ground, and diverted the course of the Oxus River through the ruins. The fabled cities of Bukhara, Samarkand, and Herat, as well as dozens of lesser cities were plundered and destroyed. We are told that of Herat's estimated 1,600,000 civil and military inhabitants, only 40 survived. More than 1,700,000 were slaughtered in Nishapur, and fewer than 100 of the 700,000 people of Merv survived. Though the numbers are probably high the magnitude of the disaster remains inescapable. A century later the renowned Muslim traveler Ibn Battuta found many of these cities still in ruins, and Toynbee suggests that the historic decline of the Muslim world began in part because of the Mongol destruction of the irrigation infrastructure and its subsequent impact on agriculture which never recovered. A thousand miles west, Baghdad with all that it represented and contained, was doomed.

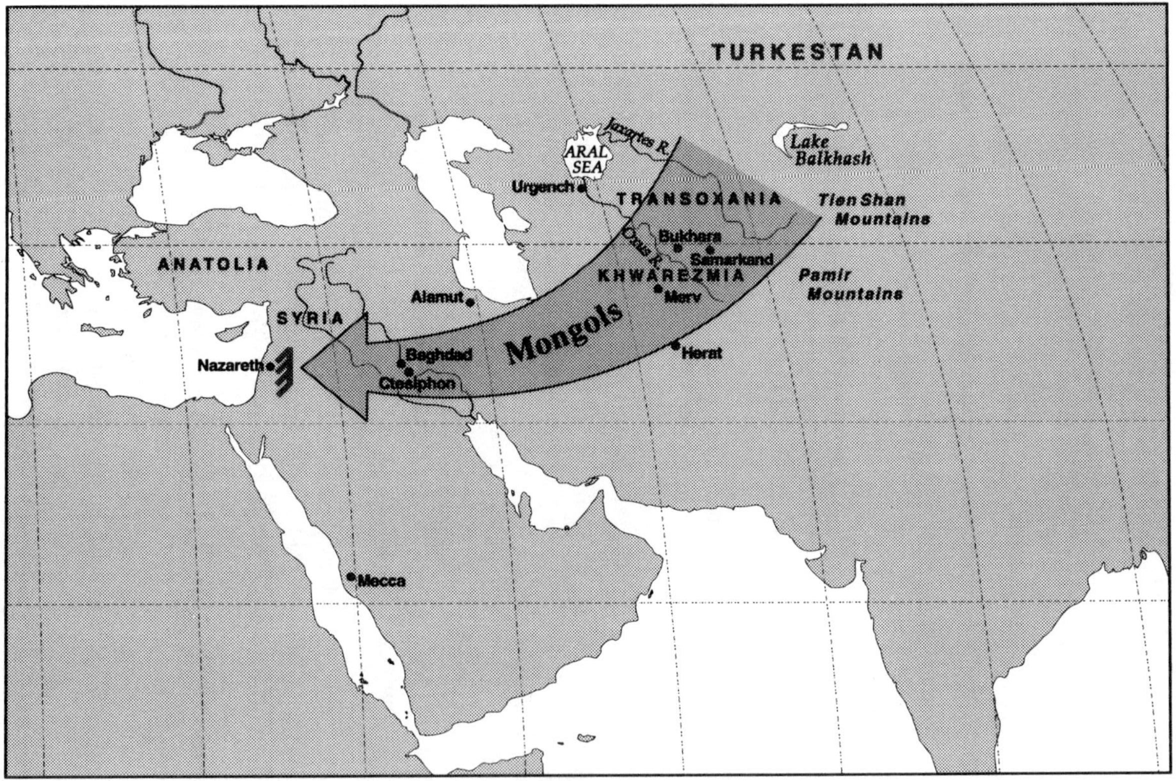

Map 3
Mongol Sweep into the Muslim World

The suicidal trigger was arrogance. In 1218, Jenghis Khan, busy directing his Mongol armies against north China, wished to secure stable and peaceful relations with the empires on his western flank. To that end he sent a trade delegation to the Sultan of Khwārezmia. One of the Sultan's provincial governors detained and executed the members of the delegation claiming that they were spies. When the Khan learned of this he sent an emissary to the Sultan demanding that the governor be surrendered to the Mongols. The Sultan had the emissary beheaded; after all, who were these nomadic savages in some vague place far beyond Lake Balkhash and the Tien Shan mountains to demand anything from a splendid and anointed leader of the world of Islām.

The Sultan made a bad call. When Jenghis Khan heard of the fate of his second mission he promptly wheeled his armies westward, and commenced an onslaught without precedent in human history, an onslaught that would change the history of the entire eastern Muslim world, and Russia as well; and, according to some historians, so weaken the Muslims that Western Europe could begin its 700 year ascendent over the Middle East.

The first major Mongol assault began in 1219, and affected mainly the eastern marches of the Muslim empire with the results just noted. The second assault came 37 years later when Jenghis' grandson Hulagu attacked and occupied much of Persia, Iraq, Anatolia, and Syria. Baghdad and 800,000 of its inhabitants were destroyed in 1258, not only ending the long impotent Abbāsid Caliphate, but terminating the institution of the caliphate as well. The Seljuk Sultanate survived in slow contraction and disintegration as vassals of the Mongols for another half century, then disappeared from history.

Mongol expansion westward through the Muslim world to the Atlantic appeared to be unstoppable. The Mamlūkes who had replaced Saladin's dynasty in Egypt were the only military force of consequence remaining before the Mongols. In what was clearly one of the decisive battles of world history, they defeated the Mongol army in Palestine, near Nazareth, at the battle of Ayn Jālūt in September 3, 1260. Had the Mongols not been stopped, the political history of Islām might have ended with the 13th century, and the Ottoman rise to power over the next 200 years replaced with who knows what.

A moment's reflection is necessary before proceeding. Our chosen focus is the Islāmic tide of history. The view is of the breadth of events. Breadth is the emphasis. Nevertheless, beyond this focus immense turbulence existed in the affairs told of here, and our story should not imply otherwise. I, like many others, have chosen to identify and speak of nominal eras such as the several Caliphates and Sultanates (Figure 2). For our purpose this is acceptable, but know that the hegemony of the Caliphs and Sultans was uniform neither in time nor in space. At any moment in any part of the empire the influence of the Caliph or Sultan could be accepted, ignored, or challenged, as indeed was virtually always the case. So, for example, even during the era of al-Mamūm's culturally wonderful Caliphate serious local challenges to his authority existed in both the east and the west. The same is true of the history of the Imāmate. Shī'ite allegiance to the Imām was as variable and fractured as that of the Sunnī to his Caliph or Sultan. With this caveat in mind we can now more comfortably proceed with observing the tide, all the while knowing that the citizen of those times, just as today, was caught up in local turbulences that very likely masked the truly, historically important events.

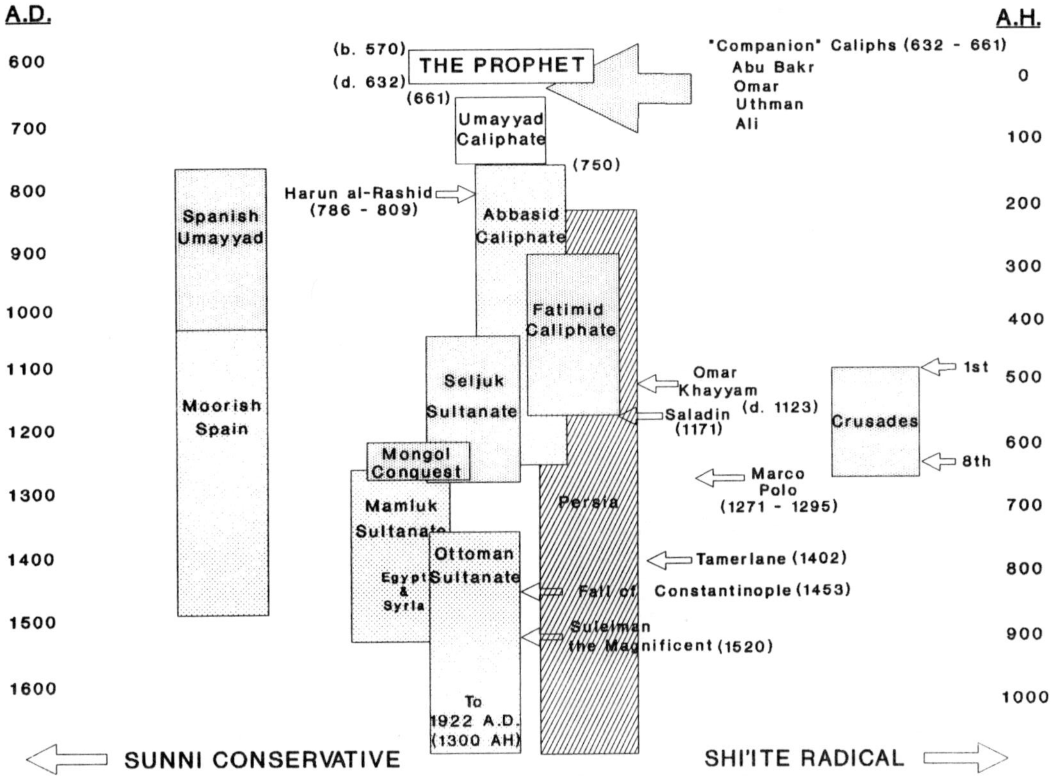

Figure 2

Simplified History of Early Islamic Power in the Middle East

Chapter 3

Transition and Decline, 14th thru 19th Centuries

Historians tell us that at the beginning of the 14th century Muslim culture would yet peak in Cairo, Damascus, Constantinople, India, and elsewhere. And yet the peak of Muslim political power was already passing. The first signs of this appeared as the Abbāsid Caliphate propelled the Umayyad fragment into Spain after 750. The Fātimid manifestation of the Sunnī/Shī'ite schism followed in 909 reaching Egypt in 969. The Turkish Seljuk dominance in Anatolia and Mesopotamia came a century and a half after that. By mid 12th century the Muslim world was so politically and religiously fragmented that it never again presented a unified threat to anybody, albeit from time to time even a fragment would get the attention of a sizable part of the civilized world. Baghdad and its cultural offerings of the ninth century were long past. That the centers of cultural glory at Cordoba, Granada, Cairo, and Damascus remained to bloom only define the separate regional elements that Islām had acquired.

In this transition era we find the world of Islām represented by six largely independent political centers; Moorish Spain, Mamluke Egypt, Ottoman Anatolia, the Arabs of central Arabia, Safavid Persia, and Mughul India. Only the Ottomans would reach the 20th century with any political influence, and by then even they were known as "The Sick Man of Europe."

Moorish Spain was sufficiently far from the Muslim center of gravity in the Middle East that events as momentous as the Mongol invasions had no effect on it at all. In the meantime, its political history took the usual course of numerous petty wars among local Muslim power centers. Eventually they were so fragmented and weakened that the persistent, centuries long external pressure from Christian Spain prevailed with the expulsion of the last Muslims in 1492. It is somehow quite poignant that the slow political decline of Muslim Spain coincided with such wonderful cultural gifts to the future as the art and architecture of the Alhambra Palace in Granada. It's not possible to travel in southern California today without being immersed in the architectural echoes of this gift.

The Mamlukes of Egypt had their day expanding and contracting westward along the North African coast, eastward into Syria and Mesopotamia, and south along the Red Sea coast of Arabia. Cairo took its turn as a brilliant center of Muslim learning and culture in the early 14th century, but eventually Mamluke decay and Ottoman power ended the Mamluke Sultanate in 1517. Largely ignored, except as a source of tax revenues, Egypt rapidly declined under Ottoman rule into what some have called the "Egyptian Dark Age" where a neglected infrastructure precipitated famine and pestilence. Egypt recovered somewhat, but for almost two centuries was to remain virtually anonymous until the arrival of Napoleon Bonaparte with his soldiers and scholars in 1798.

The Bedouin Arabs of the Nejd in Arabia had completely withdrawn from events at the center of what they had created. When Persian cultural influence overwhelmed the Arabs at the beginning of the Abbāsid Caliphate they returned to the desert and their old ways as abruptly as they had appeared. What they left behind was a world they had forever changed. In their desert anonymity, which was to last more than a thousand years, they would create and incubate relatively little of importance for our times except for the religious reform movement begun by Muhammad ibn Abd al-Wahhāb in the 18th century. Ibn Abd al-Wahhāb was troubled by the inevitable evolution of belief away from the original practices of early Islām, and especially troubled by the liberal interpretations of the Qur'ān and the *Hadīth* made by the Sūfī mystics. His essentially conservative teachings attracted one of the desert princes, Ibn Sa'ūd, who converted along with all of central Arabia in the latter 1700s. As events unfolded after World War I, Ibn Sa'ūd's descendants formed the nation of Saudi Arabia where the conservatism of Wahhābi Islām is still very much in evidence.

Persia has from the beginning been the great problem for Islām and the Arabs. Though its Sasanid rulers were weakened by internal power struggles and internecine conflict with Byzantium at the time of first contact with its Arab conquerors, and the Zoroastrian state religion was becoming more oppressive, Persian culture and its traditions in art, literature, technology, commerce, and government were nevertheless exceedingly robust and vital. Of all the monarchial traditions dating from the world's earliest history, only the Persian monarchial tradition has survived, albeit not continuously, into modern times. It is a tradition of profound import testified to by its tenacious persistence in the face of immense historical change over the millennia. Even after the Arab conquest it took centuries for Islām to be thoroughly integrated into the culture, and several centuries more before its Shī'ite tradition was solidly established. However, once achieved, Persian Islām was virtually indestructible. Its strength against foreign cultural forces was no doubt due in part to its robust millennia old traditions. Its ability to survive internal stresses must be attributed in part to the time honored *taqīyah,* the Shī'ite ability to dissemble in the face of contrary power. The efficacy of the "Hidden Imām," the Twelfth Imām described in Chapter 2, was that even as secular or Sunnī rulers controlled their subject's bodies, their souls were in the hands of the Imām. Eleven hundred years after his 'withdrawal,' the 1979 revolution in Iran saw the secular ruler Shāh Muhammad Rezā Pahlavī abruptly overthrown and superseded by the Hidden Imām's deputy the Ayatollah Ruholla Khomeini. That's inner strength!

After the devastation and occupation of the Mongols in the 13th century Persia recovered to become a local power of significance under the Safavids during the 16th and 17th centuries. Thereafter it would be more or less a pawn in the struggle between the growing power and imperial ambitions of Russia, and Great Britain. The Ottomans, too, were a player of consequence in the beginning, but their control in the east began declining even before it did in the west. The Persian state settled into its approximate current boarders in the 17th century as it withdrew from Mesopotamia. However, it left behind a large Shī'ite population, especially in the south of Iraq, which we have recently seen assaulted by Iraq's currently Sunnī leadership.

Nevertheless, however independent the Arabs, Persians, and

other Muslim states, they as well as the West often continued to acknowledge the Islāmic suzerainty of the Ottoman Sultanate. For while Islām was in retreat in the Iberian Peninsula it continued to expand at the other end of the Mediterranean, where, indeed, even Constantinople finally fell to the Ottomans in 1453. Ottoman hegemony continued to spread in advancing and receding waves up into the Balkans, through Palestine, Syria and Iraq, down into western Arabia, and around the Black Sea. Yet with all this expansion it is interesting that very few new Ottoman conquests in Christendom resulted in new conversions to Islām. (In fact, apart from Bosnia and Albania conversions to Islām have been uniformly unsuccessful in Europe, even in the part of Spain where Muslim occupation lasted 800 years notwithstanding the many Spanish Muslims that were expelled or forced to convert. The contrast of failure and success between the lands north and south of the Mediterranean is striking.) So it appears that the political power and inspiration of Islām remained effective, but the spiritual power was not generally capable of supplanting Christianity, and thus left few marks on Europe. (The expansion of the Muslim religion into central, south, and southeast Asia, carrying with it a range of political power varying from almost nothing to almost everything was substantially complete by the 17th century, but largely outside of our focus here.) By 1700 the great age of the expansion of the Muslim faith was over. Further gains over the next three centuries would be very gradual, evolutionary, and quite insignificant.

The rise of the Ottoman Turks was the natural result of the power vacuum created by the Mongol subjugation of the Seljuks after 1258. Mongol control of Anatolia was gradually weakened as the remaining Seljuk satrapy disappeared toward the end of the 13th century leaving a dozen local amirs fighting among themselves. The Ottomans came from this group. Osman, who by 1356 had prevailed over his competitors in northwestern Anatolia, established a line that would last 500 years and reach the 20th century. In 1396 a successor, Bayezid I, was so successful against the Christian Balkans that he was declared Sultan, thus cementing the legitimacy of Ottoman authority in the Muslim world. Then Mehmed II achieved in 1453 what Muslim armies had been trying to do for almost 800 years; Constantinople fell, and the Byzantine Empire was no more. Thereafter, except for a hitch when

Tamerlane briefly threatened, the Ottoman Empire would be ascendent for 200 years.

Almost immediately Osman's initial success at unifying the Turks and his clear interest in the Balkans began to get the attention of some western European powers. In less than 50 years they were more than attentive, they were scared.

At first, the Ottomans considered attacking the fortified city of Constantinople, but like their predecessors, found it too strong. Accordingly, they bypassed it and began a 200 year envelopment of Greece and the Balkan states. Serbia, Bulgaria, and Wallachia (part of modern Romania) were either directly occupied or persuaded to become vassals of the Ottoman Sultan.

A common element of such vassalage frequently required the local rulers to send their sons to live and be educated at the Ottoman court, more or less as high level hostages. The young Vlad III (Tepes) of Wallachia, born in 1431, the same year Joan of Arc was burned at the stake, passed his teenage years in Murad II's court as a companion to the future Mehmed II. His abilities were recognized, and he was promised his father's throne when the time arrived. However, his patriotism wasn't seduced, for when he returned to Wallachia and became ruler he was among the most vigorous in asserting independence of the Ottomans. (Alas, patriotism, as so often the case, did not preclude cruelty; "Tepes" means "The Impaler," and Vlad III is known in history by his given name, Dracula.)

When Suleiman the Magnificent became Sultan in 1520 the real world was entirely different from the Euro-centered one most of us learned about in high school history. We learned that the early 16th century was the age of discovery; the globe was being mapped, conquistadors were marching through the Americas subduing hostile natives and sending their golden booty back to Europe on hundreds of galleons braving storms, pirates, and privateers. The Pope had divided the World in 1494, West for the Spanish and East for the Portuguese (the dividing line in this Treaty of Tordesillas cut north south through Brazil, thus to this day all of South America west of 49 degrees speaks Spanish leaving only Brazil speaking Portuguese). Henry VIII swaggered in England, and Titian painted an equally satisfied Francis I in France. The Renaissance ebbed, and the Reformation gained momentum. Petty scheming for advantage among western European monarchs was magnified for us into life and death struggles for national survival. In central and eastern Europe the Hapsburgs, capitaled in Vienna, often seemed to be a kind of buffer between Europe and a vaguely acknowledged Muslim world.

They were a buffer indeed, and twice almost became the property of the Ottomans. Arguably the most powerful state in the world, Suleiman's Ottoman Empire had within living memory finished off forever the eternal and venerable Byzantine Empire. His rule completely embraced the countries of the Mediterranean basin except for Italy, France, and Spain. With his navy in the eastern Mediterranean, and mastery of the Middle East, he controlled commerce to the Orient, a matter which played no small part in Western Europe's decision to go a'sailing elsewhere, and thus make all those profitable discoveries. Suleiman's dominions also extended far south into Egypt, to the holy cities in Arabia, to Yemen, Syria, and Iraq, all of Anatolia, he virtually owned the Black Sea (to the chagrin of Russia), and of course, the Balkans.

Suleiman's struggle against the Hapsburgs for control of Hungary was successful enough to cause Europe to consider aiding the Hapsburgs, a strange situation since most of the time Europe was resisting them. It was, however, a classical situation, one we have recently seen again in the Persian Gulf where United States aid has alternated between Iran and Iraq. The problem was, and is, a matter of balance, to aid one enemy just enough to distract the other enemy, without making either one uncomfortably powerful. In the 1530s both France and the Ottomans were playing the same game with the Hapsburgs as the United States played in the 1980s with Iran and Iraq. In both cases the balancing act led to unexpected results. The Ottomans inadvertently set themselves up for immense future grief by granting the French, a nation they considered rather insignificant, certain commercial and legal rights within the Empire in what is known as the Capitulations of 1536 and 1569.

The court of Suleiman the Magnificent was the most splendid of its time, indeed, it was Europeans who gave him the sobriquet, "The Magnificent," for he was known in the Muslim world as "The Lawgiver." Through his vigorous building program he transformed Byzantine Constantinople into the Turkish city familiar to us as Istanbul (and though it wasn't officially renamed "Istanbul" until 1930 it was called such by Muslims from the 13th century. The word derives from a corruption of the Greek, *is tin poli,* meaning "to the city.") He was known and highly regarded for the exceptional competence of the administrators, lawyers, architects, and writers brought to his court or otherwise supported. Suleiman's treasury was fat with plunder and well sustained with revenues from his vast empire. His army was exceptionally good and included elite units of Janissaries, the soldier slaves drafted in their youth from conquered Christian lands, educated as Muslims, and trained as the most highly professional soldiers in the world. Nevertheless, they weren't invincible.

Twice Ottoman armies stood at the gates of Vienna, but Vienna was saved, in 1529 by circumstances totally beyond its control, and in 1683 by an early sign of change developing deep in the tide of history. The salvation of 1529 lay in that old bugaboo of military operations, logistics. Unlike the highly mobile Mongols who were adept at living off the land and hearty enough to easily withstand European winters, Ottoman armies required a substantial supply base, which for reasons of efficient distribution elsewhere in the Empire had to be 800 miles away in Constantinople. This supply line was just too long to sustain, and their siege had to be lifted as winter approached.

The second siege of Vienna in the summer of 1683 failed because Ottoman resources were overextended, their military technology was now significantly behind Europe's, and their leadership, with few exceptions, had degenerated into habitual incompetence. Also, the Powers in Europe were individually stronger than ever before, and they were beginning to act in concert against the Muslim threat. The Ottoman withdrawal in the autumn signaled a slow decline from which they would never recover.

Why do the mighty fall? When a State is at its peak of power and influence it seems inconceivable to its people that it could ever be otherwise. Nevertheless, it can happen, and did happen to the Ottomans beginning in the 16th and 17th centuries. As the 18th century developed the decline became apparent to all but the Turks. It wasn't until the 19th century that the Ottomans admitted their own weakness, but by then it was too late. Several attempts were made to recover, but the 20th century and the storm of World War I swept them up and blew away forever the empty shell of Empire. One could say that the Ottomans fell from a combination of political, economic, and cultural failures, or more generally, from the degenerating influence of high civilization as Ibn Khaldūn's analysis might suggest.

Politically, the focus of all power on the Sultan worked as long as the Sultans were competent. However, many were not. As with any group of humans, the corrupting influences of absolute power and unrestrained comfort had their inevitable impact. Bad leaders made for bad policy which made for imperial decline. The Sultans also distrusted everyone around them; not only was the execution of officials frequent, but even the brothers of a new Sultan were put to the sword to eliminate possible competition. In such a milieu it's scarcely surprising that the interests of the Empire were not always foremost on the minds of the bureaucracy. As time passed the Sultans began to delegate more power to their Grand Viziers. An independent bureaucracy developed that further divided authority between the Sultan, and the Grand Vizier and his staff. Corruption and intrigue flourished. Political adroitness instead of administrative ability determined success, and as the central government weakened, power dispersed to the elites in the provinces. The bottom line was a fatal lowering of the quality of government.

Economically, the Ottoman Empire was essentially a military State. The focus on the military led naturally to a certain contempt for all that was otherwise, thereby leading to a gradual neglect of agriculture, commerce, and industry. It became an exporter of raw materials and an importer of manufactured goods. This led to inflation exacerbated by the debasement of Ottoman silver currency by the enormous influx of gold and silver from the Americas. At the same time trade was bypassing the traditional routes to the Orient through Ottoman lands and following the Portuguese passage south around Africa.

Demographically, the population swelled beyond what the economy could support. This led to all sorts of problems including a serious increase of banditry by army veterans unable to find work.

Culturally, the Ottomans were trapped in the subtle and

profound prison of Islām's natural conservatism. And as their Sultanate aged into comfortable power they no longer looked outward for ideas; Islām and their own importance was all that was of any value. The West and its technology passed them by, and they would not change, or even acknowledge the need for change. Printing, for example, was largely forbidden in the Empire until the early 18th century. The Sultans were almost totally ignorant of the West until Napoleon stepped ashore in Egypt in 1798.

We shall now pass over 300 years of mutual jostling by the states of the Middle East, of the ebb and flow of events pushing borders one way then another, of the numerous conflicts between the Sunnī Ottomans and Shī'ite Persians, between Muslim states and Russia, and between Muslim states and the growingly dominant French and British Empires. We pass on, merely acknowledging that though French troops left Egypt after only three years, the seductive impression left by French (Western) culture would remain to confuse and haunt Muslims in Egypt into the 20th century. We must pass over most of this because the detail of events would too easily obscure the already identified tides of Muslim decline that are pushing us into the 21st century. Let us then, pick up the story in the next chapter, on the eve of the 20th century.

Chapter 4

The Twentieth Century

"Rule, Britannia! Britannia, rule the waves." This 1740 verse by James Thomson could have been the epitaph of the Ottoman Empire, and indeed, of the independence of every Muslim state on earth. By 1890 Britannia ruled the waves, and effectively, much of the Middle East as well, with a little help from the French, and an occasional distraction from Russia.

The growing economic power of Europe, especially in England and France, coupled with rapid advances in technology, fueled imperial ambition in the 18th and 19th centuries that swamped a Muslim world too conservative to keep up. The lands of the Middle East were becoming objects to be manipulated or pushed aside if insufficiently accommodating to the particular European power appearing at their doorstep; they were pawns in someone else's chess game.

In 1798 Napoleon landed in Egypt hoping to strangle England's lifeline to her empire in India. An unsuccessful expedition from Egypt into Syria and Lord Nelson's appearance off Alexandria persuaded him of the futility of his design, and he withdrew to France. This episode was the first clear strike of the death knell of the Ottoman Empire and Muslim influence in the world. It focused European interest on an area largely overlooked up to that time; scholars discovered the antiquities of ancient Egypt at the same time politicians and merchants started considering a canal at Suez and acquiring bases and treaties in the Persian Gulf area. All of this was taking place through a century that saw three wars between the Ottomans and Russia, while pressures increased in the Balkans gradually forcing Ottoman withdrawal from Greece, Wallachia, Bosnia, Herzegovina, Serbia, and Bulgaria. In 1875 the

Ottomans went bankrupt, followed seven months later by the bankruptcy of Egypt. A weak economy and second rate technology could not support the inflated Empire; it shrank and was kept from collapsing only because England and France wanted it as a buffer to the ambitious Russians.

The exploitation of Egypt and other Middle Eastern countries by England and France was greatly facilitated by what at the time appeared to be harmless concessions to European business interests. The Portuguese, Russians, French, and English had at one time or another, beginning in 1453 with the Venetians and Genoese, obtained commercial concessions, called Capitulations, from the Ottomans that gave extraterritorial privileges to their nationals living and working within the Empire. These privileges included exemption from Ottoman laws and taxes. The eventual problem was that ambitious Europeans could exploit Ottoman resources and yet keep the created wealth for themselves. Over time this led to effective foreign control of the Ottoman economy and the bankruptcy we have noted. This humiliating control by foreigners was also an important element in inspiring nationalistic elements in Egypt and Turkey.

As if things weren't going bad-enough for the Muslims of the Middle East, a new current was added to the tide. Zionism was a movement working for the establishment of a national homeland in Palestine for the Jewish people. It originated in the 16th and 17th centuries, but became significant only at the end of the 19th century when a combination of growing anti-Semitism, especially in Russia, Ottoman weakness, and European strength and imperial ambition made the concept potentially viable. The time was right.

Theodore Herzl, an Austrian journalist, had been thinking about a homeland for the Jews for years. He originally believed that Jews would be best served by assimilation, but when he saw that this wasn't going to happen he organized the first Zionist Congress in 1897 to discuss a homeland. It was about this time that the movement became political and was beginning to be called "Zionism." The West looked favorably upon Herzl's ideas even if the Ottoman Sultan did not.

A small Jewish immigration into Palestine was beginning with the financial assistance of the Rothschild family. The immigration rate picked up with the continuing encouragement of the succeeding

Zionist Congresses and another pogrom in Russia in 1905, but the opening of World War I momentarily stopped it.

Eventually, however, the war turned out to be more of an impetus than barrier to the Zionist cause. Three war related factors pushed this tide; the final collapse of Ottoman Empire, thus removing any organized political resistance to Zionist plans; the full maturation of British and French imperial power where they felt complete freedom to dispose of the Middle East as they saw fit and without much appeal to the wishes of indigenous populations (as nicely exemplified by the Sykes-Picot Agreement, about which more in a moment); and the need of the Allied Powers to bring as much force against the Central Powers as possible.

The British Cabinet was working very hard to get both the United States and Russia into the war. Russia had been in, but the trauma of the first years of combat and its own political disarray had greatly weakened its effort. It was thought that the population of influential Jews in both countries might move their respective governments to join the war effort if the British government supported creation of a Jewish homeland. Accordingly, on November 2, 1917, the British foreign secretary Arthur Balfour sent a letter containing what has ever since been known as the Balfour Declaration to Lord Rothschild, a leader of the British Jewish community, saying, "His Majesty's Government view with favour the establishment in Palestine of a national home for the Jewish people, and will use their best endeavours to facilitate the achievement of this object, it being clearly understood that nothing shall be done which may prejudice the civil and religious rights of existing non-Jewish communities in Palestine ..." The proviso for protection of the "rights of... non-Jewish communities" seems naive at best, and the Palestinian population was anxious from the start.

It's not difficult to understand why Arab and Middle Eastern nations today look at the West with a very cautious eye at best. The Balfour Declaration was an arbitrary action taken unilaterally by Great Britain and supported by her allies. It was clearly motivated by self interest, albeit a self interest with some altruistic intent and which seems to have been rather a cut or two above the usual dealings in international affairs of the time. Nevertheless, the Arabs saw this Declaration and its subsequent execution as a betrayal of their rights by the Allied Powers. But the Balfour Declaration was only the first half of a double barrel blast at the Middle East, and there was no mitigating element in the second shot.

While Britain and France were trying to coordinate their Middle Eastern war efforts without unduly upsetting each other they came to an agreement in 1916 through their ministers Messrs. Sykes and Picot identifying their mutually acceptable 'spheres of influence.' This is a polite diplomatic term for dividing up the pie. The pie was the Middle East, the French slice was Lebanon and Syria, and the British slice was Iraq and Transjordan. Palestine was to be nominally under international control, but the British were to have the Mandate. Since no one could see a use for the interior of Arabia, it was left to the Arabians, who, unbeknownst to all, were sitting atop a large part of the world's oil reserves. While enlisting the Arabian leader Sharif Hussein to help fight the Turks the English and French encouraged the misconception that after the war the Middle East north of Arabia would be his to rule. The details of the Sykes-Picot Agreement were kept secret. When they later became known the Arabs felt that once again they had been betrayed by the West. This double dealing is depicted in the very entertaining 1962 film, "Lawrence of Arabia."

Wars have often been the signal of historic change, but rarely in history has a war marked so pervasive and deep a change in the world as World War I and its organic sequel, World War II. The end of World War I marked the end of one age and the inchoate beginning of another. The tragic slaughter and profound disillusionment of the West in World War I, and the awesome and horrendous technologies produced by World War II are important, but not key to the issues we are following here. In our focus we find that at the end of World War I the Ottoman Empire is no more, and that the "Young Turk" faction is coming to power and fundamentally changing the political and social structure of the nation now called "Turkey." We find that out of the geopolitical machinations of France and Britain a swarm of nations have been created in the Middle East. We find that not only are the seeds of the nation of Israel well planted, but thoroughly nourished by the Balfour Declaration and the subsequent League of Nations Mandate. And we also find that a new mineral treasure has been discovered that will become the most significant determinant of Middle Eastern history for at least a century to come.

Oil! The full fruition of the industrial revolution would be unthinkable without it, and without it the dominance of the West might be problematic. It's literally and figuratively the essential fuel of our industrial and commercial engine. Its impact has been felt by everyone from the common man in his daily affairs to heads of state planning the destiny of empires. Oil changed the world beginning with its first commercial production in Romania in 1857 and in the United States in 1859. Petroleum energy, with a respectful nod to coal, transformed transportation, heating, and lighting that otherwise hadn't changed significantly in thousands of years. The "horseless carriage" was horseless because of the efficiency, safety, and economy of the petroleum driven internal combustion engine. And wisely anticipating the inevitable, the British navy at the end of the 19th century decided to convert from coal fired steam propulsion to petroleum and thereby double the efficiency of its warships.

However, the British Admiralty was not keen on relying for its life blood on the United States or Russia, who between them accounted for 90% of the world's production at the time. Fortunately, there was an alternative. (Indeed, without this alternative one might argue that the British navy's supremacy through the next half century might never have occurred. Think of what that would have meant in the two world wars!) The British had been sniffing around the Middle East for possible sources of petroleum that they could control. A 1901 oil concession with Persia finally paid off in 1908 when oil was discovered in the south western part of the country. Oil has since been discovered in Iraq in 1927, in Saudi Arabia in 1936, in Kuwait in 1938, and in the several minor countries of the Gulf region so that by the last decade of the 20th century more than 20% of the worlds petroleum production and 66% of its reserves are located there.

The Middle Eastern oil discoveries of the 1930s didn't become fully developed until after World War II in the 1950s and 1960s. The immense wealth and political leverage that came with the development of this resource were synergistically coupled to the fall of Western imperialism and the rise of Muslim nationalism.

Turkey, representing the core of the old Ottoman Empire, was the first of the Muslim states to break free of the West. Paradoxically, it managed to do this by becoming the most western-like of any Muslim country. It also had the advantage of half a millennium of imperial tradition; it was accustomed to ruling itself, not in being ruled by others as was the case in much of the rest of the Muslim world. Moreover, it was preparing for independence well before the Ottoman collapse. In the late 19th century an educated middle class of professionals developed that was dissatisfied with Ottoman incompetence and corruption, and eventually forced the Sultan to institute reforms. The organizational force, as has so often been the case in modern times, was the young military officer corps. This active revolutionary group, known as the Young Turks, had among its members Mustafa Kemal who later was responsible for the transformation of the wreckage of the Ottoman Empire into the modern Turkish state we know today. His program, carried out in the 1920s and 1930s, was breathtaking in its scope, and amazing in its success; he replaced the Arabic alphabet and script with Roman, abolished the Sultanate, closed traditional religious schools, abolished polygamy, instituted secular Western legal codes, abolished or discouraged Muslim forms of dress including the fez and veiling of women, introduced the Gregorian calendar, and ordered all families to have family names as in the West. In accordance with this last act he took the family name of Atatürk, meaning "Chief Turk" or "Father of Turks," and has since been known as Kemal Atatürk.

In the process of building a new nation he managed to keep the West, and Russia, from partitioning Turkey, to keep Greece from biting off a big chunk after World War I, and later to make peace with Greece. Thereafter, Turkey continued to modernize and provide a modest prosperity for its people even without the benefit of the prodigious oil revenues accruing to many of its Muslim neighbors. Of particular note is the fact that Turkey has remained Muslim and politically stable, apparently in spite of its substantial Westernization.

The period between the World Wars saw Western influence dissolve and nationalist movements arise in many of the Muslim proto-states of the Middle East. The League of Nations legitimized the Balfour Declaration and the Sykes-Picot Agreement by substituting Mandate responsibility for the *fait accompli*. From the legal point of view this meant that Britain and France could do what they pleased with the Middle East, but neither the League of Nations, nor Britain, nor France had bothered much to consult with the Middle Eastern peoples. Moreover, it seemed that nothing could control the tide of Jewish immigration into Palestine, especially after Adolph Hitler came to power in Germany in 1933. This situation was a continual problem for Great Britain, because the Palestinians were becoming increasingly unhappy with the virtual invasion of their land by the Jews, and the Jews were reacting to Palestinian resistance.

Muslims had had enough of European control and began fighting for their independence. Although clearly outgunned, they were troublesome enough to eventually succeed. Sometimes complete independence came in stages with an initial period of rule by a monarch appointed by the British, as in Iraq and Egypt. Sometimes there was a long, relatively trouble free incubation before independence, as in Transjordan. Sometimes there was a long drawn out guerilla war as in Algeria. In any case, independence came, even if only nominally at first, to Egypt in 1922, Iraq and Saudi Arabia in 1932, Lebanon in 1943, Syria and Transjordan in 1946, Libya in 1951, Morocco and Tunisia in 1956, and Algeria in 1962.

The British with their hands full trying to keep control of events in Palestine finally concluded in 1937 that their Mandate was unworkable, and accordingly proposed the partition of Palestine. No one liked this idea, neither the Palestinians nor the Jews who both objected to the proposed boundaries. The hiatus of World War II only made matters more acute, and the British eventually turned the whole mess over to the United Nations which also recommended partition in 1947. The State of Israel was proclaimed in 1948 when the British left.

The independence of Transjordan went relatively smoothly given its physical linkage to Palestine. Transjordan annexed the West Bank (territory extending 15 to 30 miles west from the Jordan River and the Dead Sea) including half of Jerusalem, and changed its name in 1949 to the Hashemite Kingdom of Jordan, or less formally, simply Jordan. During this time national identities developed in Jordan and the other new states of the Middle East contributing to an isolation of the Palestinians from their Arab brothers. After the United Nations permitted the State of Israel to come into existence in 1948 a brief war ratified the fact, and the Palestinians were more or less left adrift without consistent, long term support for their cause in the international community. Whatever was done by the other Arab states was rather episodic and often motivated by interests other than Palestinian welfare.

Meanwhile, Persia hadn't escaped the attention of Britain, or Russia either. Disastrous dynastic conflicts had virtually destroyed the Persian infrastructure and greatly facilitated the growth of foreign influence. Throughout the 19th century British and Russian commercial and geopolitical interests manipulated Persia with the same disregard for its people as the British and French were to do later with the rest of the Middle East through the instruments of the Balfour Declaration and the Sykes-Picot Agreement. Britain viewed Persia as a very important communications link with its Indian Empire, and later as a vital source of oil. Russia, always interested in expanding its Empire, especially where there was potential for access to warm water ports, looked longingly at Persia (and not for the last time) at the same time that it forced its commercial interests. Persia's savage and corrupt Qājār dynasty was overthrown by a military coup in 1921 bringing Rezā Khān to power. Rezā Khān was supported by the Majles (National Consultative Assembly, ie., Parliament) and the mullahs (religious leaders) who, fearing Persia might be transformed into a secular state with the rejection of many Muslim customs as was happening in Turkey under Kamel Atatürk, eagerly supported his assumption of the crown in 1925 and he became Shāh Rezā Pahlavī.

In the process of making the country more independent and competitive with the West, the Shāh eventually affected many of the reforms that the mullahs feared; Iran (for so it had been officially renamed in 1935) became a secular state with education and law largely taken out of the hands of the mullahs, Western style higher education was greatly encouraged, and traditional Muslim dress codes for women were abolished. The seeds of the 1979 revolution were sewn. There remained only the nourishment of tyranny and poverty to bring them to fruition.

Apart from the matters already mentioned, that is, the dissolution of Western imperialism, the emergence of independent

Muslim states in the Middle East, and the final precipitation of the State of Israel, World War II (and its precursor World War I) had one more effect of interest in our story, the emergence of the two Super Powers the United States and the Soviet Union. The Soviet Union, and Russia before it, had long been interested in Persia and the Ottoman Empire. At first Russia feared the Ottomans, then as the power balance changed, she coveted parts of its territories. At one time or another Russia aspired to control of the Bosporus and other lands in Turkey and around the Black Sea. She wanted the Caucasus between the Black and Caspian Seas and even fancied making Iran one of her Soviet Socialist Republics after World War I. As history has shown, it never got the Bosporus, nor an Iranian Soviet Socialist Republic, though it did get the Caucasus and the northern side of the Black Sea.

By the 1950s the Cold War between the Super Powers had provided a wonderful tool for the Middle Eastern states where they effectively sold their 'friendship' to the side that would pay the most. Egypt is easily the best example of this, switching sides twice. Unfortunately, most of the 'payment' was more often than not in the form of military rather than economic aid. This led to numerous dictatorships that have had very little indigenous support, and as these dictatorships were almost invariably secular, they have led to some interesting balancing acts. The Muslim clergy was unhappy with the sell out to evil, non-Muslim powers, and the people in their poverty were unhappier still. To control this, Middle Eastern leaders professed Islām while building their armies. Money better spent in relieving poverty instead flooded into a military armed to the teeth and never quite sated with a quantity of arms far beyond any reasonable defense need.

We can identify at least three major currents in the tide of Middle Eastern history for the last half of the 20th century; the overwhelming influence of oil, the turbulence of secular Arab nationalism and Islāmic fundamentalism, and the growing presence of the state of Israel in Palestine.

The political leverage accruing to the producers of Middle Eastern oil became apparent in 1974 when they embargoed shipments to countries that supported Israel in the 1973 Yom Kippur War. The impact on the West, and necessarily on the rest of the world, was as shocking as it was dreadful. Gasoline prices tripled and quadrupled triggering an inflation that soon demonstrated to the Arab oil producers that such embargoes were two edge swords that drove their costs up as rapidly as their income. Moreover, Western energy conservation and competition from non-Arab oil producers eventually weakened this source of political leverage. Nevertheless, both the Middle East and the West discovered a new power balance in the world.

In 1959, in the first coordinated effort to control their destiny, Arab oil producers joined with other oil states around the world to form the Organization of Petroleum Exporting Countries (OPEC). Although it worked for awhile, independent market forces eventually took charge and left many extravagantly spending Middle Eastern states deeply in debt. Nevertheless, oil revenues gave formerly backward and vulnerable countries such as Saudi Arabia, Iraq, and Iran prodigious sums of money. By 1975, Saudi Arabia, the richest of the oil states, had financial reserves exceeding those of the United States and Japan combined. Some countries used this new wealth to buy weapons, some to improve their infrastructure, and some to invest in the West. In most cases, however, the new national wealth did not relieve the common man of his misery. This fostered a profound discontent in the masses and contributed to a hate of the West that was viewed at once as their exploiter and seducer. Since the United States was the West's quintessential materialist state it often became the special target of Muslim rage.

Nowhere is this better illustrated than in Iran. The United States in its Super Power struggle with the Soviet Union encouraged the Shāh to develop his armed forces, a plan the Shāh was pleased to pursue for his own ends as well. The Shāh's reinstallation of the imperial throne with its accompanying splendor and extravagance left the Iranian masses very much as they had been before the discovery of oil, only this time their misery was easily contrasted with the crown's comfort. And this time there was also the alternate promise of Islāmic fundamentalism. The Ayatollah Ruhollah Khomeini successfully returned from exile in Paris by promising to abolish the monarchy, restore the jurisdiction of the *Sharī'a,* and lift the oppression of the common folk. He persuaded all groups and classes in Iran to support the revolution by avoiding mention for the time being of his most radical plans.

Even the middle and upper classes, often with only a nominal interest in their Muslim heritage, detested the Shāh's regime, especially because of his hated and feared secret police, the SAVAK. The net result was a quick revolution in 1979 that largely rejected the West and established the new Islāmic Republic of Iran. Once in power Khomeini no longer needed the support of the "moderates" and proceeded to dump them as his extreme programs were enacted. The whole process is reminiscent of the Abbāsid overthrow of the Umayyads 12 centuries earlier where the essentially Sunnī Abbāsids used the Shī'ites and *mawālī* to come to power, then rejected them when no longer needed.

The essence of our story, and the final focus of events, is the Middle East's reassertion of its independence with its more or less successful escape from Western domination through its development of secular Arab nationalism, and almost simultaneously, Islāmic fundamentalism.

The 20th-century Muslim world has been seriously experimenting with secular, nationalistic government, most notably in Turkey, Egypt, Syria, Iran, and Iraq. As we have seen, Turkey appears to have been the most successful in this experiment, Iran, an utter failure. For the others, the jury is still out.

The most important Arab figure in secular nationalism was Gamal Abdul Nasser, born in 1918, undisputed leader of Egypt in 1954 two years after a group of army officers ousted King Farouk, and president of Egypt until his death from a heart attack in 1970. Nasser's promise was to restore dignity to the Arab people who had so often been humiliated by the West. His success and failure was directly tied to the degree that he could accomplish this. In the beginning he was very successful. He nationalized the Suez Canal Company and succeeded in thwarting a combined British, French, and Israeli invasion to regain control of it. He demonstrated that his own technicians were well capable of operating and maintaining the Canal in the face of Western assertions that Egypt was incapable of such a feat. He charted an independent course for Egypt in the Super Power struggle choosing on his own who he would befriend. When the West wasn't forthcoming with aid to build the High Dam across the Nile at Aswan he turned to the Soviet Union with better results. He inspired other secular Arab nationalists in Libya, Syria and Iraq, and briefly formed the United

Arab Republic in union with Syria. Generally speaking, he was very highly regarded by the peoples of the Middle East. However, this was not the case with Arab rulers, most of whom either distrusted his ambition, or were unsympathetic with his governmental secularism.

Nasser's success turned to failure as he coped with the growing hostility of other Arab leaders and with the dissolution of the United Arab Republic in 1961, but the most serious blow to his image came with the humiliating defeat in the 1967 Six Day War with Israel. Israel's attack destroyed Egypt's air force even before it could get off the ground, while Egypt's army was man-handled and ejected from the Gaza Strip and the Sinai. Israel achieved all this while dueling Syria and Jordan with its other hand.

Egypt's economy began to deteriorate, partly because it was unable to support the cost of Nasser's military establishment. The combination of humiliation and hard times seriously reduced his popular appeal and eventually brought into question the efficacy, if not the very legitimacy, of secular Arab nationalism. King Feisal in Saudi Arabia was constantly raising this question and assumed more influence in the Arab world as Nasser's influence declined. To compromise the concept further, other secular Arab nationalist states such as Iraq, Syria, and Libya were becoming one man dictatorships that were highly oppressive and exploitive of their peoples. The time was ripe for the reemergence of a force that had lain largely dormant for several centuries, a force that called to the very soul and being of every Arab and Iranian, fundamentalist Islām.

"Islāmic fundamentalism" is not a well defined term. Rather like "Christian fundamentalism" it embraces many sects and points of view. Its one common meaning and point of view is the notion that the true Islāmic life and government is attainable only through a return to the basics of Islām. Islāmic fundamentalism is manifest by Sunnīs and Shī'ites alike, and it would be a profound mistake to identify it only with the vocal Shī'ites of revolutionary Iran. On the other hand one Muslim's notion of fundamentalism can be another's idea of apostasy or conservative corruption (witness Iran's hostile view of the traditionally conservative and Islāmically pure Wahhābī Saudi Arabia.)

Islāmic fundamentalists are chiefly concerned with the corrupting effect the West has had on Muslim culture. While some elements of Western culture such as science and economic power have impressed them, they completely reject the seamier parts that have weakened what they believe is their historically superior civilization. They are particularly troubled by alcoholism, bars, theaters, night clubs, movies, drugs, the unrestricted public mixing of sexes, women's Western dress, pornography, materialism, secularism, and the absence of compassion that has permeated their culture from the West. Muslims feel the threat of Western ideas even in their own homes where emancipation of women and the growing disrespect of their children for their elders has had no precedent. But the essence of the inspiration for 20th-century, Islāmic fundamentalism is the profound sense of humiliation felt by virtually all Muslims at the hands of the West. Muslims, or at least the Arabs and Iranians, have very long memories. The absence of the glories of Islām and the decline of its brilliant cultural centers in Baghdad, Cairo, Damascus, Constantinople, Cordoba, and Granada are difficult to accept, especially in the face of the belief that it is somehow unnatural and sinful for misbelievers to rule believers.

Islāmic fundamentalism isn't a new phenomenon. It was manifest in the very earliest days of Islām with the Sunnī/Shī'ite schism and notably later with the birth of the Fātimid Caliphate. It has periodically reoccurred numerous times in Muslim history driven by whatever the current source of dissatisfaction with the status quo. In recent times the tide of Western domination and humiliation has been that source of dissatisfaction.

Although there were calls for a return to basics in the 19th century, the first important manifestation of Islāmic fundamentalism in modern times came in 1928 with the founding of the Muslim Brotherhood in Egypt. It and its offshoots remain today arguably the most important, though by no means the only, fundamentalist movement in the Middle East. At its founding it stressed moral and social reforms in response to a deep feeling that the reason the West was so successful in dominating the Muslim world was because Muslims had corrupted their practice of Islām, and God was punishing them for it. Islām had to rid itself of Western secularist ideas and decadence. The West was at once

God's cause and effect of Muslim decline. However, the call for moral and social reforms was found to be insufficient, so in 1939 the Muslim Brotherhood became political declaring that, "Islām is a comprehensive, self evolving system; it is the ultimate path of life in all spheres. Islām emanates from, and is based on, two fundamental sources, the Qur'ān and the *Hadīth*. Islām is applicable to all times and all places." By 1946 the Egyptian Muslim Brotherhood had grown to a million members and sympathizers. They found inspiration to resist British influence and control in the Brotherhood slogan, "The Qur'ān is our constitution, the Prophet is our guide; death for the glory of Allāh is our greatest ambition." (It is worth noting, however, that though, "... death for the glory of Allāh is our greatest ambition," extremely few of the million chose to exercise that ambition. Clearly, this slogan is a characteristically Arab utterance with its completely predictable exaggeration, passion, and verbal substitution for the act.)

The Muslim Brotherhood now exists in all Arab states, either under its own name or another name assumed for political reasons; in Algeria it's The People of the Call (*Ahl al Daawa*), in Tunisia it's The Islāmic Party (*Hizb al Islāmi*), in the Gulf States it's The Society of Social Reform, and in Syria it's the Young Men of Muhammad (*Shabab Muhammad*). In addition to these Sunnī groups we find numerous others, some Sunnī and some Shī'ite such as the Party of God in Lebanon (*Hezbollāh*), Algeria's Islāmic Salvation Front, the Islāmic Resistance Movement (*Hamas*), and the Islāmic Jīhad, both in the occupied territories in Palestine, and the Shī'ite Amal. Scholars estimate that there are perhaps several hundred fundamentalist organizations across the Muslim world with most in the Middle East. Analysis, however, suggests that total membership is less than two or three million representing less than 0.3 percent globally and less than one per hundred in the Middle East.

Fundamentalist or secularist viewpoints are very rarely manifest when confronted with political expediency. In 1980 the Islāmic Republic of Iran put aside its fundamentalist principles (to the consternation of the fundamentalist Syrian Islāmic Front) and made alliance with secularist Syria against secularist Iraq. At the same time, elements of the fundamentalist Muslim Brotherhood chose to support secularist Iraq. There is a principle operative here that should not be forgotten: In a confrontation between religion and politics, politics will almost always prevail in the short run, and always prevail in the long run.

The variety of fundamentalist movements suggests the wide appeal of fundamentalism among the Arabs and other Middle Eastern Muslims. That fundamentalism hasn't swept the day may only be due to the disorganization and fragmentation of the phenomenon, for contrary to some Western fears, it is largely local with local peculiarities often at odds with similar movements elsewhere. Islāmic fundamentalism is no where near being a coherent, universal Muslim movement. In those countries where the secularist governments are oppressive and have failed to cope with the problems of poverty Islāmic fundamentalism is finding ample opportunity for expansion. The impoverishment and humiliation of Nasser's Egypt by the 1967 Arab/Israeli War nicely illustrates a secular Arab nationalist failure and the subsequent opportunities provided to the fundamentalists. Probably only the very strong repressive measures by Nasser kept the country from turning into an Islāmic State. A similar dynamic has existed in Hafaz al Assad's Syria and Saddam Hussein's Iraq where secular states are maintained through the oppressive instrument of their respective Baath Parties and secret police. How long these three states can survive this tide is an open question.

Where conditions have been less stressful for Muslim peoples there seems to be less inclination to support fundamentalist movements. Turkey and Saudi Arabia come immediately to mind. However, perhaps the best example was the occasion of Anwar Sadat's assassination in Egypt by members of the Al-Jīhad movement in 1981. The plotters had expected the assassination to trigger a general uprising and demand for the installation of an Islāmic State to be ruled under the *Sharī'a*. No uprising took place, probably because the country's prosperity had improved from the low of the late 1960s.

As the last decade of the 20th century opened, a continuing confusion of historically minor storms distracted one from a sense of the tide upon which these events are riding. Sunnī, secularist Iraq and Shī'ite, fundamentalist Iran concluded a decade long war in 1988 that exhausted both and left neither a winner. United States involvement here and with the never ending Palestine/Israel crises continued to provoke Muslim rage and provide much of the inspiration for Islāmic fundamentalist movements. Iraq had no sooner freed itself from its war with Iran than it invaded Kuwait in the summer of 1990. World intervention through the United Nations with largely United States and European military forces ejected the Iraqis, but there was substantial ambivalence in the Muslim community about involving the West to settle what many saw as an essentially Arab problem, oil interests notwithstanding. As the immediate threat from Iraq to other Arab states subsided after the Gulf War their support of the UN (the United States) in this matter palpably cooled, and Arabs once again began experiencing the feeling of humiliation at the hands of the West.

A historic social experiment is taking place in the Middle East. The experimental question is, can one of the world's great religions become an effective social instrument in the 21st century in the face of a contrary secular world culture of immense vitality and historic success? No question of greater historic scope has ever been asked. The dynamics before us involve Western culture that is rational, scientific, and successful beyond the wildest dreams of any person born before the 18th century. Its success is so seductive because it not only promises everything, it delivers, but there's a price. For all the material wonders obtained, the West has had to surrender a certain spiritual strength. An accommodation was reached with religion where 'spheres of interest' were recognized that effectively pushed aside religion and transformed the West into a largely secular culture. The physical benefits from this deal have been wonderful, but the spiritual and social consequences have left a void acknowledged even by the West. Because we have lived with this evolving transformation for several centuries, we have had some opportunity to adjust, however incompletely. Moreover, this is a product of our own heritage. On the other hand, the Islāmic culture of the Middle East has had no such comforting buffer. For 1,400 years they have emersed themselves in a way of life that submits to the harmonious needs of the inner man, that introspectively accommodates its world with no need to conquer the enveloping physical environment. The supremacy of religion and its complete identity with the political and social milieu answered all questions with a reverent bow to the past. The Qur'ān

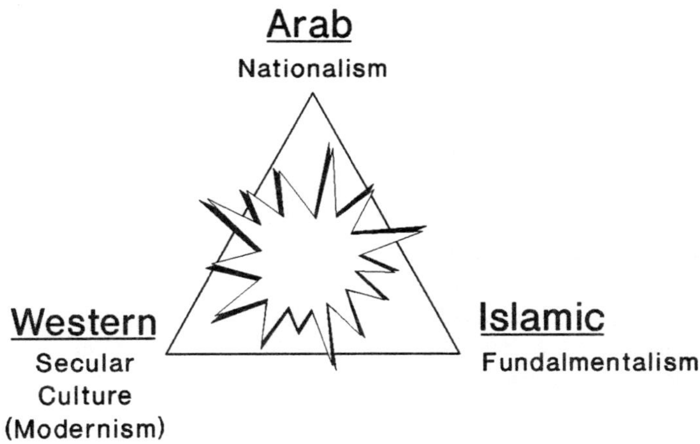

Figure 3

Muslim Middle East Triangle of Internal Conflict

and *Hadīth* provided all that was necessary and sufficient. A comforting situation indeed, but materially it led nowhere, at least in the long run. With the passage of time Western material advances superseded the splendid cultural achievements of Islām. The Muslim world ignored this until it was forced on them. The cultural shock was devastating. All at once Muslims were second class citizens in their own lands. Even their glory was second class. What was worse still was that the Muslims themselves wanted the Western material well being, but they had neither the heritage nor the leisure to facilitate the change. By the mid 20th century the Muslim Middle East had found that the material benefits of the West's culture were not worth the spiritual price they were paying, so they began a search for something more acceptable. Different styles of nationalism, socialism, secularism, and most recently, Islāmic fundamentalism, have been, and are being tried. Some attempts have failed dismally, others have had some limited success, and others have yet to bear fruit.

If one word were to be used to describe the modern inspiration for Islāmic fundamentalism it would be "humiliation." To the degree that secular Middle Eastern nationalists fail to deal with the humiliation experienced by their peoples at the hands of the West, to that degree will Muslims turn to fundamentalism, especially if they are economically depressed as well. Where pride is evident, as for example in Turkey, and in Nasser's Egypt before the 1967 war, and to some degree in Saudi Arabia, there one will find less interest in Islāmic fundamentalist revival. Accordingly, whatever the West does to humiliate and take the pride of the people (who incidently are not necessarily synonymous with the government) that will increase the success of Islāmic fundamentalism.

Of course other factors are operative here as well, after all, Saudi Arabia can scarcely be viewed as a state humiliated at the hands of the West, and yet it too has been struggling with militant fundamentalists. Clearly economic discontent and simple political ambition are also operating. Indeed, the symbiosis of economic discontent, political ambition, and militant fundamentalism often mix cause and effect into an indistinguishable mess.

The main contest at the moment is between secular nationalism, a Western import, and Islāmic fundamentalism. Algeria, Egypt, Iraq, and Syria are the principal nations under varying degrees of siege in the 1990s. In Algeria the legacy of French culture has left an influential secular residue which allowed the country to experiment with socialism. After more than a decade of peace and relative prosperity, partly based on oil revenues, the experiment failed as economic problems fueled popular discontent. With the legalization of the Islāmic Salvation Front in 1989 the fundamentalists got 55% of the vote in local elections and later gathered 49% of the vote in the first round of national elections in 1991. The military seized power and cancelled the second round of elections when it was evident that the fundamentalists would win and carry out their stated plan to transform the country into an Islāmic state under the rule of the *Sharī'a*. The Islāmic Salvation Front was resuppressed as rioting and assassination followed, including the assassination of Muhammad Boudiaf, the popular president of the ruling council. It's clear that some form of compromise will have to take place, or that the economic condition of the people will have to be improved if Algeria is to avoid a repeat of Iran's experience. Of course, oppression can work for awhile as well.

Egypt has been on an economic roller coaster for decades, and accordingly has experienced several episodes of zealous fundamentalist activity. As in many other Muslim states when the economy is good the fundamentalists have had only marginal success at best. However, the opportunities during a failing economy are another matter, and become so attractive and unique that the militants have recently taken to attacking the sources of what ever prosperity Egypt can muster. Their rationale effectively says 'if you won't support fundamentalism because you're comfortable, we'll remove the comfort, and then you can reconsider.' In an acute sense of Egyptian economic vulnerability the fundamentalists began attacks on foreign tourists and businesses in 1992 hoping to deprive the country of this very important revenue source. The plan succeeded admirably causing a substantial decline in tourism. The government responded with increased security for tourists and a very aggressive assault on the fundamentalist infrastructure. Although tourism seems to be recovering, how this will eventually turn out remains in question. It's worth noting that in recent centuries Egypt has often felt rather closer to the West than many Arab countries except possibly

Algeria and Lebanon. Accordingly, appeals to throw off the humiliation of Western dominance have been less effective. The fact that the fundamentalists have had to resort to attacking the economy and thereby appeal to the material interests of the populace rather than their Islāmic heritage supports this view, although the militants maintain that the attacks are to drive out the corrupting Western travelers and businesses. The argument is unpersuasive, albeit typical of radical groups fanatically committed to maintaining ideological purity.

At the moment (1996) Iraq and Syria have things pretty much under control, principally because of the strength of their individual dictatorships. Control and repression are so thorough that no significant organized alternatives to their secular governments are possible. However, because their regimes are totally personal, the withdrawal or death of Hafaz al Assad in Syria or Saddam Hussein in Iraq will surely open unique opportunities for the fundamentalists. The struggle for political supremacy will be between the military and the fundamentalists, since there is no other organized interest in either country. Iraq with its Shī'ite majority and proximity to Iran is perhaps more vulnerable than Syria, but both are largely Arab countries heir to the discontents with the West we have previously mentioned, and accordingly subject to the persuasive arguments to reject the West and return to the original purity of early Islām.

United States foreign policy in the face of Iran/Iraq interactions has been driven in part by a wish to prevent any fusion of interests, and especially any political fusion. The possibility of a unification of Iran and Iraq, with their immense oil resources and consequent geopolitical leverage, is more than we can accept. Accordingly, the Sunnī/Shī'ite and secular/fundamentalist divisions between them suits the U.S. just fine. U.S. foreign policy during the Iran/Iraq War of the 1980s illustrates this. We wanted neither side to win.

Another American anxiety was that a unified state under Islāmic fundamentalist rule would exert a powerful force in the Muslim world toward a further expansion of fundamentalism and its anti-Western and especially anti-U.S. attitude. However, the fear of a local or general pan-Islāmic union seems unrealistic in light of the history we have summarized in the past four chapters.

Not only is there little in the one and a half millennia history of Islām to suggest any potential success in such an undertaking, but the very character of the Arab militates against it. Their tendency to fratricidal conflict is of legendary proportions, with intergroup animosities inversely proportional to the distance between them. The success of the seventh-century Arabs can only be partly attributed to the cohesion of the *Umma* defined by Muhammad. The attraction of plunder for these nomads must have been equally as powerful. Today, the only plundering probable would be of each other, a not very auspicious prospect for the beginning of a new empire.

A successful long term union of Iran and Iraq, of Persians and Arabs, seems equally problematic given not only their character differences but their deep religious differences as well. Recall that in this context the forces of union are the various fundamentalist movements, and that by their very nature they are extreme and essentially inflexible. We have seen for example how fundamentalist groups will fight each other, even siding with secular forces to prevail. This rigidity guarantees an unstable union.

Nevertheless, there are centripetal forces at work too. For example, the irritation of the state of Israel in their midst is an exquisitely painful reminder of Arab and Muslim humiliation at the hands of the West. And their are memories of the glorious Muslim civilization which existed while the infidel Europeans were barely more than a bunch of forest savages to the northwest. Moreover, there is the shared wealth of oil revenues that in spite of the disproportionate distribution to the rich, still helps the poor majority if only by providing jobs now and then. The *zakāt* is after all one of the five pillars of Islām and becomes as efficacious now with oil revenues as it was in the seventh century with booty from Persia and Byzantium.

As we look at the Middle East today we see a storm of local interests carried upon the tide of emerging self assertion. This self assertion first became prominent in the half century following World War I with the political independence of Muslim states from Algeria to Iran. Economic independence, to the degree that any state can have it, came largely with the growing world importance of oil, especially after the oil embargo and OPEC increases in

1973 and 1974. But the deepest and most profound self assertion is occurring right now as Middle Eastern countries battle to achieve cultural independence. Some countries feel the need for it more than others; Turkey's self image seems to have been secure enough that they have unabashedly taken Western cultural elements to themselves ranging from the Roman alphabet to Western style clothing, names, and laws. And they have accomplished this without renouncing their Muslim faith. At the other extreme Iran, through the vehicle of Islāmic fundamentalism, has ferociously rejected Western culture. Other Islāmic fundamentalist movements, rarely connected to the Iranians, have been pressing their programs elsewhere, all in service of breaking the Western cultural influence in their lands.

It is natural for a people to want to be free of any bondage. For the Muslim peoples of the Middle East the Western bondage that remains is like a fine scarf; in some ways it's comfortable and pleasing to wear, but restricting and humiliating, because its long tag is held by someone else, and it could be fatal if tightened like a noose. The task and challenge is how to slip the noose. Some, like the fundamentalists, say destroy it. Others say transform it; keep the comfortable, stylish qualities and dispose of the dangerous, restricting, and humiliating elements. Each approach has its promise and problems.

The central problem for Middle Eastern Muslims is to reconcile the rigid and archaic practices of early Islām with the exigencies of today, for if one accepts the view of the Iranian scholar, Dr. Husain Nasar, as many Muslims do, that, "For Muslims, Islām is not one of the elements of life; Islām is life itself" they are faced with a paradox, because the genuinely Islāmic life advocated by fundamentalists is incompatible with the modern world. It reflects an unrealistically nostalgic view of the past as well as a mischievously naive view of the present. This incompatibility has been demonstrated time and time again.

In Muslim countries political expediency always prevails when confronting religious principle, a behavior of course not unique to Muslim countries, but unique in the size of the gap between practice and principle. Islām has always faced the conflicting imperatives of religious doctrine and political reality, notwithstanding its claim of universal jurisdiction in life. From the viewpoint of a religious purest, it has been an utter failure, more striking because of its claim on life. Of course Muslim scholars deny this (a common Shī'ite rational cites the Qur'ān's permission to use dissemblance in response to overwhelming political repression) but their arguments are unpersuasive and smack of sophistry. Christianity has faced this conflict too, but it's been notably more successful in the accommodation with the ready rationale of, "Render unto Caesar..."

Islāmic fundamentalism is proposing a path that cannot succeed, neither locally nor globally. Given the absolutist characteristics of the various sects there is no cohesion nor potential for cohesion among them, and as local isolates, inward looking communities will simply be washed away by the tide of history as incidently as a sand castle on the beach vanishes with the passing tide. If they choose to be flexible and compromise, they become what they had condemned. So it appears that in the broadest strokes, Islāmic fundamentalism has little long term potential for success, albeit even in its failure the short term (a few decades) could be a very trying time for the West.

The short term indeed has its perils; the 'oil card,' nuclear weapons, and the appearance of the true Mahdī, to name some obvious possibilities. As the tide of history slowly moves before our eyes these are especially worth watching.

What the Muslim Middle East does with the economic gain and political leverage of oil is a short term issue that will have the greatest long term impact. So far this immense windfall has been largely squandered in the purchase of weapons used mostly for show in an exhibition of machismo unparalleled in history. The world's proven oil reserves are about one trillion barrels, or about a 50 year supply at current consumption rates. Since the Middle East currently has virtually no other asset for generating wealth, it must look to these oil revenues as the sole source of leverage to develop the capability of competitively sustaining itself in the 21st century. It has less than the lifetime of one human being to make the transformation. Muslim children of today can expect to see their world without significant oil resources as they pass into old age. To the degree that we see the Muslim countries of the Middle East use their oil revenues to develop infrastructure like roads, railroads, airports, communications, seaports where applicable, power plants, and desalination or other water supplies, as well as education, and a suitable industrial potential, to that degree we will be seeing a historically significant tide. Otherwise, Islāmic fundamentalism, secular nationalism, or any other 'ism', will be nothing more than a passing historical storm, and the Islāmic Middle East will fail to completely reemerge and establish its full potential in history.

Oil has influence far beyond the mere revenue it brings in. Ever since the formation of OPEC and the oil embargo the West has been much more accommodating to the wishes of the oil nations than it might have been otherwise. The challenge for the Muslim oil states is to use this leverage just enough to get what it politically wants without making the West fear that its vital interests are threatened and thus provoking the West to military intervention. The Persian Gulf War was only a hint at what could happen if these larger interests are disregarded or miscalculated by future Saddam Husseins.

The issue of nuclear weapons lurks in the background as one considers the future of the Middle East. Not only has there been demonstrably active indigenous programs to develop nuclear weapons in Iraq and Iran as well as in nearby Muslim Pakistan, but the collapse of the Soviet Union has made available some of their weapons as well. The organizational chaos in the constituent states of the old Soviet Union has permitted individuals and institutions to sell nuclear materials. With their fractured lines of command and neutralized governmental authority only the most wishful thinker could believe that items as valuable, marketable, and portable as tactical nuclear weapons are not 'available' to those with the money and connections. Few in the world have more money and better connections than the Muslim Middle East. Indeed, it is said that Iran has already purchased four nuclear warheads from the Republic of Kazakhstan and as of 1993 is merely awaiting delivery. When the propitious moment arrives to announce their possession of nuclear weapons the geopolitics of the area will change forever. Whether the restraint of a "mutual assured destruction" policy would work is problematic given that only Israel is certain to possess a nuclear counterforce. Nuclear blackmail seems highly likely in the region. This of course could draw the nuclear powers into alliance with the nuclear have-nots which could either ease the threat or exacerbate it. The threat to the West itself would also be real, albeit not on the same scale as that to the Middle East. The advantage of the West in facing any nuclear threat from the Middle East is one of vastly superior supporting military technology without which a nuclear capability is severely limited. It takes rockets and very sophisticated electronics to effectively deliver nuclear weapons any distance. Of course, terroristic, clandestine delivery is also possible, but the risk of premature discovery is also higher. Moreover, to insure victory in their use against the West, a very large number of them would have to be used. The acquisition of more than a few dozen weapons by a Middle Eastern country is probably unlikely for some while yet. The threat then is not so much to the West, though there are surely many fundamentalists who would like to make it so, the threat is to the Middle Eastern Muslim world itself. The history of international relations in the area is not encouraging. This is especially discouraging to consider, given the highly emotional nature of the Arab, and the absence of that quality of diplomatic savoir-faire that tempers and adds wisdom in judgement. Cunning is not wisdom, and unlike wisdom, it is the slave of passion.

Of all of the features of Islām the concept of the Mahdī has the most potential for mischief. The first notion of the Mahdī originated with the death of one of Alī's sons in 685, and evolved into its present form with the disappearance or "occultation" of the Twelfth Imām, the "Hidden Imām" in 874. The messianic idea of the Mahdī, which means, "rightly guided one," expresses the belief that the Twelfth Imām will reappear to restore proper Islāmic belief to the world and bring on a short era of several years duration where truth, justice, and virtue will prevail before the end of the world and the Last Judgement. This is clearly of Shī'ite origin, although some Sunnī Muslims, even without Qur'ānic or reliable *Hadīth* support, have from time to time accepted the concept as well. These millennial prophets tend to appear when times are difficult for Muslims, a phenomenon also familiar to Christians. Individuals calling themselves the Mahdī have appeared regularly throughout Muslim history. Perhaps the best known in the West was shakyh Muhammad Ahmed who led an uprising in the Sudan in the 1880s against the Egyptians.

Ahmed finally occupied Khartoum and killed General Charles "Chinese" Gordon in 1885. The Mahdī died under mysterious circumstances five months later. His successor ruled till 1898 when a British force under the command of General Kitchener reoccupied the Sudan and killed him.

The mischievous element in the mahdī belief springs not only from the blatant political agenda invariably espoused by the claimant, but by the eschatology. Key is the triumph of Islām (and of course it's the Mahdī's version of Islām) as a preparation for the end of the world. Before the nuclear age such millennial expectations were harmless. It is, alas, no longer so. It takes no great imagination to foresee a passionate Muslim leader saturated with ambition and zeal for the final triumph of his religion, a leader who finds one or more nuclear weapons in his hands, who would plan to detonate them as a natural, indeed inevitable, conclusion of God's will. And lest one believes that this is too wild to ever be possible, we should remember that the deeply religious President Reagan once publicly speculated on the millennial role of the United States nuclear arsenal. In any event, the mahdī tradition is an ideal instrument for the would-be demagogue, and it would be very surprising if none appeared in the next few decades.

Chapter 5

Conclusion

My object in telling this story of more than thirteen and a half centuries of Islāmic culture has been to establish a sense of scope and importance. I have chosen to pass far above the waves and storms of detail in order to view the whole sea of history. I hope that in so doing the view will more easily reveal the tides that otherwise would be lost in the splash of momentary events. This is relatively easy to do for eras past, but not so easy as we approach our own time. That's the challenge, not only for the writer, but for the reader as you try to interpret the events before your eyes, what is significant and relevant to the tide and what is not.

Part of the challenge is in putting aside the personally convenient units of time, the days, weeks, months, and years so appropriate in our everyday lives, and adjust to the longer perspective of history. Our personal perspective inevitably leads to a distortion of perceived time. Just as visual perspective presents objects at a distance as foreshortened, so too events and processes centuries and millennia distant appear to span but an instant. Yet an historical "instant" is often a decade, a lifetime, or more. For example, the Arab conquests after Muhammad's death are invariably described as "rapid," "explosive," "amazingly speedy," "in an astonishingly short time," etc., yet the military conquest of some countries took decades and the conquest of the future Muslim world took centuries. The Islamization of the conquered territories took centuries more. From our perspective it happened in a flash; from the perspective of an Arab merchant in the year 650 it encompassed his lifetime, and those of his children, grandchildren, and several generations thereafter. Clearly, the time unit of history's clock exceeds a lifetime. One says of historic changes that they happened in the past or will happen in the distant future. The beat is often subtle and difficult to identify for historic change of the present. We feel that our world, at least the world close to us, is more or less permanent. And so it has been for all men; imagine for a moment how permanent one's world must have appeared to the citizen of Hārūn al-Rashīd's Baghdad in 800, or of Suleiman the Magnificent's Constantinople in 1530 (or indeed, of Bill Clinton's Washington, D.C. in the mid 1990s). Yet the tide of history measured in decades and centuries reveals them as passing affairs sometimes leaving no trace whatsoever. Even the original site of Hārūn al-Rashīd's Baghdad, its glories only 11 centuries distant, but the victim of neglect and sack, is problematic since the modern city has been rebuilt many times and in slightly different places from the original. So we must take the longer view. Somewhat paradoxically, however, we are often powerfully persuaded that the event of the moment is the decisive hinge upon which all the rest of history will swing, but that's rarely so.

The longer view, alas, is not facilitated by modern news media. The media, and especially the electronic news media, necessarily seek the sensation of the moment for their own natural, commercial ends and contribute little to an understanding of the true import of events. They are shackled by their very efficiency at presenting events of the moment; they are, after all, instruments of immediacy projecting momentary images. Their ephemeral presentations lack the permanence of a book, and lacking this permanence give us no sustained occasion to reflect and question their substance. There can be no dialog with the TV. On the other hand, one is reminded of the possibilities of leisurely contemplation in Gibbon's farewell after his epic consideration of the decline and fall of the Roman Empire; "It was among the ruins of the Capitol that I first conceived the idea of a work which has amused and exercised near twenty years of my life, and which, however inadequate to my wishes, I finally deliver to the curiosity and candor of the public." Here we have the opportunity to converse with Gibbon for 2,000 pages, and for a lifetime if we wish. In our "Tides of History..." we share this taste in temporal sweep. Hopefully, these pages will be an instrument of perspective and an occasion for reflection not found with electronic media.

Part 1: Conclusion

What good is it to have a longer view of events; after all, what possible interest can we have in events that unfold over decades and centuries, especially as most often we will not have been witness to their beginning, nor will we be there for the denouement. Part of the answer is of course that there is no beginning or end. History is process. Given the turbulence of events on the personal scale, the historic significance of the moment is all but lost, and surely can be of no consolation to the victims of the moment. For those who are not the victims but the beneficiaries of the moment, most often the personal rewards accruing from the process consume our fullest attention. Victim or beneficiary, the reality in which we are immersed is usually not only primary but exclusive, or to quote a black friend of the author's remarking on reality in the inner city, "Reality is a brick on the side of the head." However, where circumstances permit for the less acutely involved observer, the long term perspective of history can be satisfying and fruitful. We can discriminate and discover the links between the moment and the tide. Then the long term view will lend meaning and understanding to events often seeming to be meaningless. Understanding leads to wisdom which leads to serenity and wise action. So we can observe, identify, and understand the process and tide of history in which we are afloat, and out of that perhaps even derive some pleasure.

Although the issue of what drives history is largely beyond the scope of this work it is perhaps worth a moment to look back over our story and see if it gives us any hints in the matter. Clearly some individuals can have truly historic impact; arguably, Muhammad was the single most influential individual in human history. Alī, Abu al-Abbās, Suleiman the Magnificent, and Gamal Abdul Nasser have had lesser impacts. On the other hand, it is difficult to separate cause and effect in distinguishing whether the individual effects history or history carries the individual. In his masterpiece, "War and Peace," Tolstoy asserts that the 'great man' theory of history is all wrong and that Napoleon, the Russian Emperor, and the several Field Marshals and Generals were merely riding the tide of events and were simply carrying out what history allowed and ordained for them. From this point of view Muhammad didn't do anything that wasn't already in the cards. The time was right and had Muhammad not come along, someone else would have. Remember the weakness and instability of the Byzantine and Sasanid worlds, remember the restless energy of the Bedouin tribes, and remember the prevalence of the *hanīfs*, the monotheistic and messianic prophets of the time. So who can say for certain which is cause and which is effect; perhaps each is both, and like the theory of chaos in modern science, extremely small differences in either can grow to profoundly influence everything, what the scientist calls sensitive dependence on initial conditions.

Elements of human behavior are a rich and never ending source of 'initial conditions' to seed history. Greed and the thirst for power, individually and collectively, are important in our story. The Muslim surge out of Arabia in the seventh century was largely driven by the promise of plunder with religion probably nothing more than a organizing rationale. Had the potential for plunder not been there it is unthinkable that the Bedouin would have attacked anyone other than themselves. With the successes of Islām came an endless succession of power struggles. Mu'āwiyah against Alī to begin the Umayyad Caliphate, al-Abbās against Marwān II to begin the Abbāsid Caliphate, Shī'ites against Sunnīs, Shī'ites against Shī'ites, and Assassins against everybody, all driven by a thirst for power only sometimes rationalized by the zeal of religion. The Ottoman Sultans often schemed to get power, and having achieved the sultanate, just as often murdered their brothers and other potential pretenders to keep it. Modern times have seen no change as we view the careers of Shāh Rezā Pahlavī in Iran, Khadafy in Libya, Assad in Syria, Saddam Hussein in Iraq, and any number of Islāmic fundamentalist leaders with specious claims of high moral purpose. Human ego should always be suspect when its bearer finds himself prominent. Idealism is also important, but it rapidly loses its soul as it attains power (and before we become too smug in pointing to the failings of our Muslim brothers we should humble ourselves by recalling that the West has produced at least as many examples of comparable human behavior).

The need for certitude appears to be a powerful force in history, and nowhere is that need more inflexibly manifest than in religion. For Muslims it is believe or die, or sometimes, pay more taxes. Shi'ites and fundamentalists are notably less inclined to grant the tax option. Spiritual certitude can sometimes be serenely directed inward as has occurred in Sufī mysticism, but then it can be aggressively directed outward as is often reflected in the Qu'rān. The world's great literature is full of this struggle between freedom and certitude; Dostoyevsky's *The Brothers Karamazov*, and the Bible's Genesis to name but two.

Certainly other factors appear to have an influence on history as well. Climate, geography, and geology to name a few. Climate seems to have conspired with geology to produce in the Arabs a hardy people hungry for expansion and plunder and capable of more sacrifice than their adversaries. Ibn Khaldūn recognizes this toughness as a necessary ingredient for the foundation of empire. He remarks that nomads are the source peoples for civilization, the renewing blood stock that rejuvenates a decayed civilization. The geographical inheritance of the Ottomans astride the outlet of the Black Sea and gateway to Asia contributed to their power for centuries and in the end prolonged their life because of the West's desire to have a buffer and blockade against the Russians. Of course the oil blessings of geology on the Middle East are too obvious and profound to need further remark.

Karl Marx's economic theory of history, though terribly one dimensional, receives partial support from the ebb and flow of Muslim fundamentalism. We have observed that material well being has frequently smoothed the highs of passionate advocacy and the lows of despair. No example of this principal better illustrates the matter than the secular-fundamentalist power duet in Turkey and Egypt during the last half of the 20th century. If the people are comfortable then the niceties of religious dogma are overlooked and the humiliations of foreign influence are grating but bearable. Accordingly, materialism is the Great Curse of revivalist Islām, and its not difficult to see why given its seductive nature. Is it any wonder then that the mullahs of the Islāmic Republic of Iran call the most materialist of nations, the United States of America, the "Great Satan."

So now we consider the breadth of our vision and try to identify the tide of evolving Islām through the centuries and millennia, into the present, and beyond. The first matter that strikes us is that not only is there more than one tide but that there is a hierarchy of tides, tides and currents at all historic scales pulling events this way and that, with lesser flows sometimes apparently overwhelming the deepest currents. A similar and illustrative situation exists in astronomy where we observe the motions of the Earth in space. The Earth has a motion affected by the gravitational force of the Sun even while it also circles with the Sun around the center of our galaxy. At the same time the Earth, the Sun, and our galaxy flow with other galaxies of the local galactic group in a direction and speed determined by the overall gravitational force of a super cluster of galaxies in this part of the Universe. And finally, there is the dominant motion of the Earth in the Universe defined by the expansion of the Universe as a whole.

The ebb and flow of the deepest, most profound historic tide is perhaps reflected in the rise and fall of civilizations and cultures. This cycle has been repeated since the dawn of history; civilizations emerge, grow, prosper, become decadent, weaken, and are assaulted and overwhelmed by barbarian conquest. Thereafter, these simpler, more energetic conquerors evolve, become more sophisticated, are seduced by the comforts of what they have taken, prosper, become decadent, weaken, and themselves fall victim to new conquerors. In recent centuries the cycle has changed somewhat because of the impact of technology. The fall of the civilization is delayed because of the immense advantage given it by technology. However, eventually enough technology seeps through to the "barbarians" to catalyze their superior *esprit de corps* and energy, all while the civilization continues to weaken, so that the "barbarians" prevail after all. (Neither "civilization" nor "barbarian" should be literally restricted; "civilization" can mean almost any dominant political group, and "barbarian" any out group. For example, the Chinese word for barbarian also means foreigner.) Ibn Khaldūn says that this process is unstoppable, while both Thomas Jefferson and Mao Zedong have explicitly noted the benefits of periodic or continuous revolution to maintain a civilization and preclude its decadence.

We have seen this largest of cycles with the rise and decline of the Muslim world itself, especially in the Middle East. We have also seen it within the history of Islām with the rise and fall of the Umayyads, the Abbāsids, the Seljuks, and the Ottomans. And we have seen it in an interplay of the strong and weak as interests grow powerful, faction develops, and unity is fractured; the

Sunnī/Shī'ite schism, the multiple simultaneous caliphates, the simultaneous sultanate/caliphate, nationalism, and the secular/fundamentalist dynamic.

The tide of Islāmic revival is now before us. Is it a major flow, a lesser flow, or minor turbulence? Is the question answerable? Perhaps the answer lies with understanding how the Islāmic revival in the Middle East articulates with the tides just noted. Does it follow naturally from the great cycles of history or from the internal rhythms of Islām?

Muslim dominance in an area typically (but not always) began with military conquest followed immediately by political control. Later, sometimes in only a few years, sometimes taking as long as several centuries, followed religious conversion of the population and after that cultural integration. There are numerous exceptions to this, but generally speaking this was the order of things. Today the Islāmic Middle East is fearful of a reversal, of losing its religious and cultural integrity. The question is, is the Islāmic revival a tidelet or The Tide of history? Will this revival, even if successful, have major and century-long lasting consequences for the world as did the first appearance of Islām in the seventh century? Will it significantly reverse the centuries-long decline of Middle Eastern influence in world affairs? If the answer is yes, then we are looking at The Tide.

Clearly, with the integrated socio-politico-religious quality of Islām any Islāmic revival necessarily implies for a Muslim a simultaneous cultural, religious, and political recovery as well. Is this enough to reestablish the former glory of Islām?

Perhaps the four most important events contributing to the decline of the Muslims and the Middle East are the incredible devastation from Transoxania to Iraq caused by the Mongol conquest of the 13th century, the reawakening of the West beginning in the 15th century with its subsequent commercial bypass of the Middle East around Africa to the Orient, and the West's technological development. These essentially economic events are what started the ebb of the Muslim tide. The forth major contributor was the essential conservatism of Islām.

If economic destruction and social conservatism are the villains, is Islāmic revival the necessary and sufficient condition for recovery? Not if it is in the form of Islāmic fundamentalism, for by its very nature fundamentalism preaches a return to values and mind sets that have proved inadequate in the face of Western competition.

What about oil resources? Surely this can be the instrument of economic recovery for the Middle East. Unfortunately, the historic evidence suggests that the oil wealth currently being squandered by privileged and corrupt classes, as well as on culturally useless military hardware, is unlikely to ever significantly benefit the overall health of the region. Indeed, it has every characteristic of a cancer. Without other significant resources it is imperative that oil revenues go to establish a viable economic infrastructure that will sustain the region after the oil is all gone. This is not happening, and there is no evidence that it will happen in the future. Iran and Iraq with their vast oil resources are pathetic examples of this historic mismanagement. The case of Iran is especially instructive since its major asset of oil, used largely in support of its sterile Islāmic revolutionary aims, is the lubricant of its slide to exhaustion. Ironically, here Islāmic revival is not the vehicle of resurgence but the vehicle of decline.

Of course the religious leaders of the Middle Eastern Muslim world deny this, but the facts speak for themselves. Moreover, there are other dimensions to the Islāmic revival that suggest its limits as a force for change, at least of change for the better if not for reestablishing the major tide of influence in the world.

It is instructive to note that with all the fuss being made by the religious leadership in the Muslim Middle East there is remarkably less supporting popular zeal. Of course we see the highly organized popular demonstrations from time to time with their ecstatic and angry shouts supporting whatever local cause, but for the millions upon millions of common people leading lives of simple faith in the ultimate goodness of God and struggling to capture a moment's succor here and there between birth and death there are few grand visions of empire for Islām. The Islām practiced by the masses has been called "folk" Islām. It is often remote from the public Islām presented and encouraged by the shrill Muslim leadership. It provides the cannon fodder of war, the soil and soul of the economy, and the litmus test of historic movement. And here the litmus test fails. From Turkey to Indonesia to the Sudan it reflects the majority culture. It is often hated by the mullahs for its flexibility and has been difficult to control. The striking absence of specifically Islāmic inspiration and rationalization in the Palestinian drive for freedom from Israel may very likely reflect the true position of folk Islām in the Muslim revival. Clearly, political/economic matters are substantially more important to the average Palestinian Muslim than the purity of Islāmic practice or the supremacy of Islāmic law. Imperial Islām, West hating, activist, fundamentalist, is a product of relatively few Muslims invariably committed to a political agenda so aggressively pressed that the religious element is almost invisible. This suggests that the Islāmic revival of the late 20th century is barely more than a disconnected group of national political movements led by ambitious men using the cloak of religion for their own ends. Historically, Islām was invariably used to legitimize claims to political power, the most recent of which was Saddam Hussein's public displays of piety during the Gulf War of 1990-91. But when the masses have been called upon to support such claims, especially in modern times where support of the masses has become more important, the masses have declined. We note the conspicuous absence of any will to fight for Saddam Hussein in 1991, or of a called for popular uprising with the assassination of Anwar Sadat in Egypt in 1981, as well as the failure of a popular revolt to materialize when the Grand Mosque in Mecca was taken over in November, 1979, and the absence of Iraqi Shī'ite interest in furthering Iran's struggle with Iraq. So there is a clear gap between militant Islām and folk Islām. This is not the sort of indicator one would expect were the Islāmic revival a manifestation of some deep current of history.

Indeed, the often heard wistfully angry call to return to its roots, to go back to basics may be a reaction to pure change more than anything else. Recall our own reaction when the familiar is taken away or destroyed while many new and unfamiliar things happen around us. Then consider for a moment the situation for a typical urban resident of the Middle East. At one time there were neighborhoods, quiet and familiar, into which he could retreat for comfort and respite from the world at large. The old urban design had streets based on pedestrian traffic and a hierarchy of scale from the quiet personal residential court connected to other courts by small passages leading off narrow streets that wound about finally making it to a main thoroughfare. Then the Western motor car arrived punching wide swaths of highway through the local 'quarters' invading and destroying the intimacy of the neighborhood with a flood of noise, air pollution, and congestion. And not only his neighborhood but his home is invaded as well. From a culture traditionally slow paced he is within a lifetime overwhelmed with images of change via radio then TV, his children mock the traditional authority of the father, his wife and his daughters behave in 'promiscuous' ways, his proud heritage of Muslim dominion is humiliatingly made subservient to the wills of not only non-believers but non-believers who he believes are the source of all that is decadent in the modern Muslim world. Under such culture shock, if indeed not Alvin Toffler's "future shock" who wouldn't react and reject it all and want to go back to the good old days? It is inconceivable that fundamentalist-revivalist Islām wouldn't have appeal under such circumstances. It doesn't have to be just; it doesn't have to be true; it doesn't even have to be realistically possible; it only has to give hope.

But the message of fundamentalist-revivalist Islām may have a fatal flaw, for its claims to perfection through the Qur'ān, *Hadīth,* and the *Sharī'a* are insupportable when administered by imperfect men and in a world so different from the source world of these documents. The contradiction is that to adjust to modern circumstances the fundamentalists will have to deviate from their fundamentalism and thus destroy themselves. Such altruism is not seen in history. So the fundamentalists will continue with their vision, poisoned by its unsuitability to the times, and by the inevitable corruption of imperfect men pretending to be perfect. This has happened several times before in Muslim history from local mahdist movements to the rise and fall of the 10th-century Fātimids in Egypt and the 16th-century Safavids in Persia. Accordingly, it appears that the Islāmic revival is more consonant with the internal rhythms of Islām and thus not a major tide in global history. It is possible, however, that the Middle Eastern Muslim's search for a satisfying and viable identity might be achieved by a synthesis of Western and Muslim values facilitated by Sūfī flexibility and mysticism. Similar Sūfī syntheses have worked in the past.

On the other hand, the potential of oil and nuclear weapons, suitably used by the Muslims of the Middle East, could profoundly change the picture. If oil revenues were used to create an infrastructure and commercial base that could sustain these countries after the oil runs out, and moreover, if these revenues were to be used to acquire nuclear weapons that are undoubtedly on the clandestine weapons market now, or shortly will be, then the global political and economic leverage would have every possibility of having lasting substantial consequences. In which case we would be witnessing a major tide in history.

The future is not predictable; we merely choose between guesses. As we examine the currents of history we try to pick out of contemporary events those items that most likely could signal the presence of important tides. For the reemergence of the Islāmic Middle East perhaps the appearance of a successful pan-Islamic Mahdī, the uses to which oil revenues are put, and the success or failure to acquire nuclear weapons may be three important indicators of the tide at hand.

PART II

The Rise and Fall of the Soviet Union

The tides of history are exemplified in no better way than the rise and fall of the Soviet Union. Seldom has history given us a more perfectly encapsuled story than the period from October 1917 to December 1991. Yet the fullest appreciation of the meaning of this time comes only with a sense of the preceding time. The decay and corruption of the last Tsarist regime and the confusion and disaster of World War I paved the way for Lenin and the Bolsheviks to take power and end 400 years of autocratic rule. October 1917 was as pivotal as July 1776. However, what evolved thereafter was quite a different matter. Ruthless idealism linked to a tradition of unbounded power made inevitable the travails of the Soviet people until the appearance of Mikhail Gorbachev. In the interim Soviet suffering and energy produced, not for the first time in Russian history, an amazing surge towards modernity and world power. And also not for the first time, this progress turned out to be relatively superficial and without the deep cultural integration necessary for permanence. The resulting stress in trying to equal and pass the West, combined with, ironically, the "internal contradictions" of Soviet socialism, led to the collapse of the Soviet Union in August through December 1991. Like the runner from Marathon to Athens in 490 B.C., the Soviet state ran as hard as it could and then dropped dead from the effort.

The rise and fall of the Soviet Union will be the tide of our attention for the next few chapters. However, history, like the ocean's tides, is connected and predicated upon events and influences spread over far more space and time than the immediate focus at hand. Almost everything is connected. To change metaphors for a moment, history is like a machine with a maze of intermeshing gears. No gear operates totally independently of any other. If one turns, all turn in some way or another. This interconnectedness is implicit even in Arnold Toynbee's monumental *A Study of History* where he begins by trying to identify civilizations that are totally independent of each other. He could find only 13 "autochthonous" civilizations in all of human history. So, not surprisingly, we find that to understand the rise and fall of the Soviet Union we must first get some sense of the history of its Russian precursor. This will introduce us to one of the pleasures of reading history, its fascinating interconnectedness.

We have seen ever so briefly the tide of Muslim history flow and ebb and flow. In the process we have seen several references to connections with Russia via Constantinople and the Black Sea, especially since Ottoman times. These connections will be touched on again, now from the Russian side.

We will also see how the pre-Ottoman Byzantine Empire, that frustrating barrier for so long to Muslim expansion northwest out of Anatolia, and especially its capital, Constantinople, played a profound part in shaping the Russian world view and helped to make Russians so characteristically Russian in Western eyes.

We have seen the Mongol impact on the Muslim world, but that scarcely exhausts the matter, for the Mongol conquests were of the greatest historic scope ever known. While the Muslim world south of the Caspian and Black seas was being devastated, an onslaught was also occurring to the north that eventually would find Mongol horsemen swarming over eastern Europe, burning Cracow in Poland, sacking Pest in Hungary and Zagreb in Yugoslavia, reaching the northern outskirts of Vienna, and approaching within 60 miles of Venice. Western Europe was saved, not because of its ability to resist, for no such ability existed, but because in 1241, back in Karakorum, Ogadei Khan died, and Mongol custom required his relatives and generals to return for the election of a new Khan. In the course of these conquests, almost as an incidental matter for the Mongols, Russia was wrenched from its rather fragmented preimperial status dominated by the cities of Novgorod and Kiev and made a satellite of Mongol interests for 200 years. Out of this experience emerged the Moscow hegemony and the beginning of a Tsarist Russia that would end only with Lenin's revolution in 1917. Thus the Mongol hand is revealed in establishing the Russian State as it destroyed the vitality of Muslim states to the south.

All this merely establishes the context for the matter at hand, the rise and fall of the Soviet Union. We shall discover the story of how young communist idealists effected a revolutionary movement that was surprisingly consistent with this previous historic context, and how the Russian people, in a perfectly consistent way, declined to express the required revolutionary zeal. This led to Lenin's revision of Marxist theory to fit the new situation. Lenin's revision also planted the seed for the eventual fall of the Soviet state. We will follow these developments from Lenin's ideal through his and Trotsky's surrender to the temptations of power, to Stalin's utter absence of ideals and the subsequent evolution of Soviet Russia into a world power. We will see how the seduction of power becomes complete and Lenin's bad seed bears fruit under Brezhnev's regime. Mikhail Gorbachev will try to rectify the situation with an idealism that is tragically inadequate for the job, and the forces latent in Russian history will once again become manifest and bring the socialist experiment to an end.

In the end we will review events since 1991 and try to see what sense they make and how this tide of history flows.

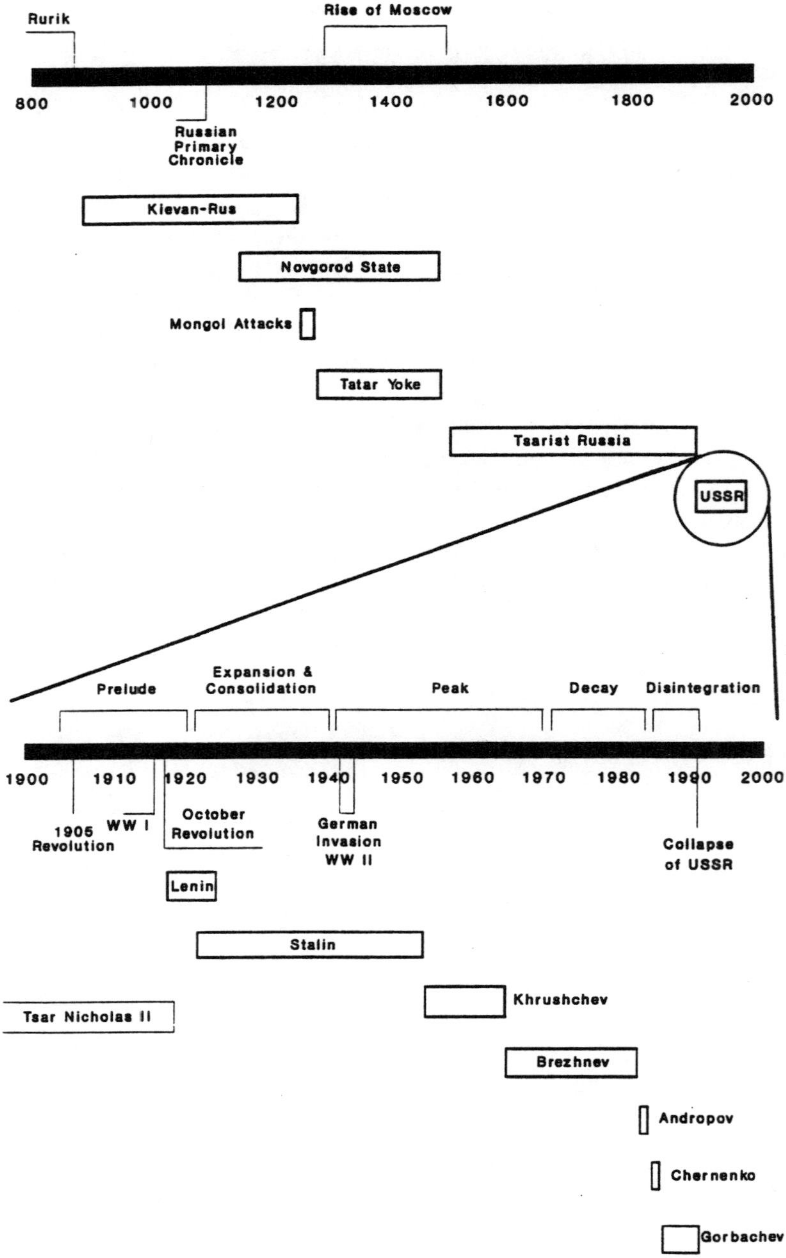

Figure 4
The Ages of Russia

Chapter 6

Out of the Mists of Time

For Russia geography is everything. This was as true 1,100 years ago as it is today. Geopoliticians today continue to remark on Russia's lack of "blue water" ports to the world's oceans and the difficulty of conducting the nation's business across the 11 time zones of its east-west vastness. Geography also ordained a largely inhospitable climate with long, bitter winters limiting agriculture and enforcing months of reduced productive activity each year. Its rivers, which have been its main arteries of communication and commerce until the coming of the railroad in the 19th century, flow largely north and south. There is no equivalent of America's Ohio and Missouri rivers reaching 2,000 miles east and west. Russian rivers not only established the internal nexus of affairs, but also determined its links with the outside world thus profoundly influencing its culture and history.

In the ninth century A.D. the most significant of these links was with the Viking traders who passed in the north from the Neva, Volkov, and Lovat rivers to the south by the Dnieper River and coasting west 400 miles across the Black Sea to the Byzantine Empire and Constantinople.

Like Arabia before Muhammad little is known about the interior of central Eurasia before the ninth century except what can be gleaned from archeological digs and an occasional remark in classical literature. From the beginning of recorded history it has been the source of endless waves of barbarians assaulting the tribes and civilizations at its fringes. From time to time between 800 B.C. and 800 A.D. some of these barbarians mingled with the locals and settled around the western edge of this *terra incognita* in the area of eastern Europe, and north of the Black Sea as far east as the Don River and the upper reaches of the Volga. (Map 4) Eventually, out of this swirling tide of peoples, a group called the East Slavs became resident in the steppe of what is modern day Ukraine. Kiev on the Dnieper River, one of the major towns in the region, anchored the area in the south. The Slavs had also pushed up into the central and northern forests south of modern Finland where the town of Novgorod provided a northern anchor. By the middle of the ninth century the Slavs had settled into a number of city states along the Neva-Dnieper axis, but raiding foreigners and conflict with each other left them with no peace, thus, as the story goes, they called in the Viking traders to rule them. It is at this time, in the year 856 A.D.. that the name Rurik first comes to our attention.

In several fascinating parallels, the leaders of Novgorod called in an outsider to take over in the face of internal discord just as the tribes of Yathrib (Medīna) had called upon Muhammad to arbitrate their differences 234 years before in Arabia. The selection may have been a natural one since the Vikings, later called the Varangians by the Russians, had begun to settle along the river trade route they had opened. Contact with the incredibly energetic Vikings possibly persuaded the Slavs that here was just the strong, unifying hand needed to get their house in order and resist the attacks of surrounding raiders as well. This is consistent with one scholar's description of the Slavic self image of the time as inert, easy going, lackadaisical, incapable of political or other initiative, and requiring stimulus from outside. The *Russian Primary Chronicle* itself, compiled in 1116 in part from a number of earlier sources, describes at least one of the Slavic tribes, the Polyanians, as, "...a meek and unwarlike people." On the other hand, the "invitation" to the Vikings may not have been as uncoerced as tradition makes out. From earliest times the Slavs were often ruled, loosely or tightly, by others. This is seen even in the word "slave," already derived from "Slav" by the ninth century. In any case, "Rurik the Viking" and his followers came to Novgorod in 856 and became the leaders of the East Slavs. This event is often cited as the beginning of the history of "Russia." Within a century the Vikings had united the warring East Slavic city states, moved their capital from Novgorod in the north to Kiev in the south, and through an occasional warlike gesture as well as profitable trade

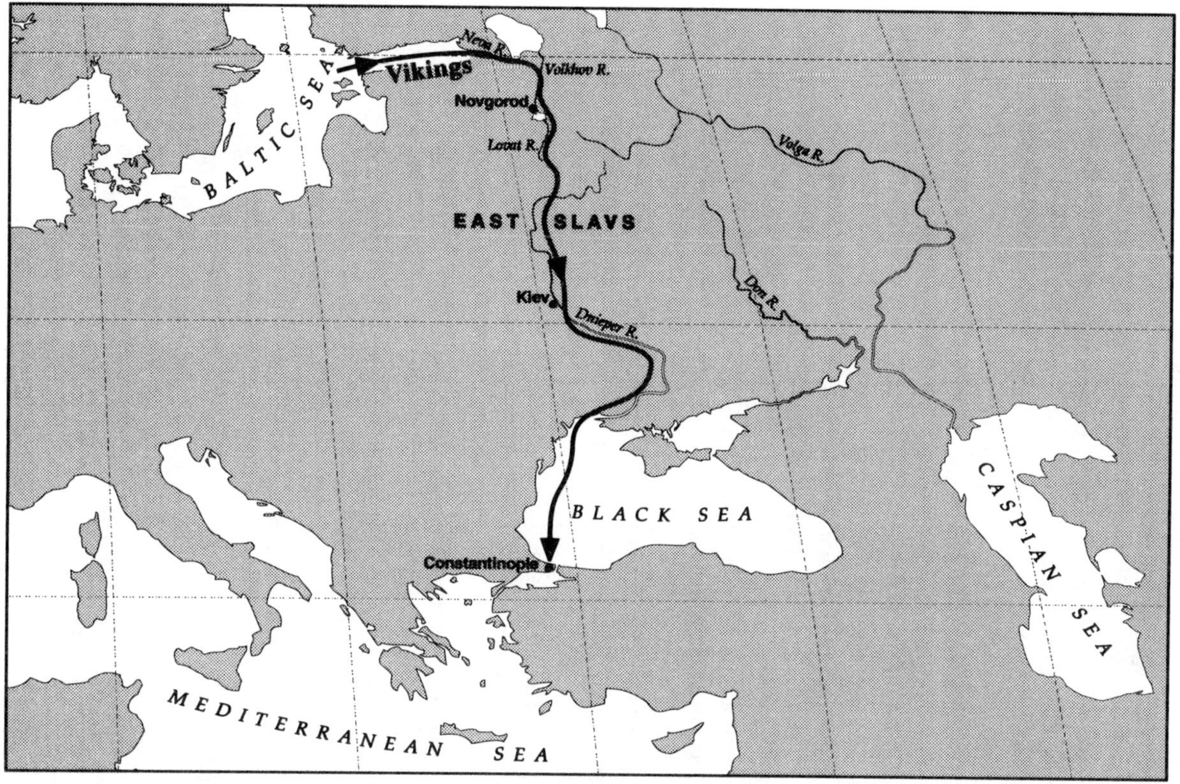

Map 4
The Dawn of Russia

with Constantinople raised the stature of Kiev to the most important center in Russia, indeed, today, we call the country of the time "Kievan Russia."

In less than two centuries the Vikings were absorbed into the Slavic culture they were asked to lead leaving behind little more than their name. "Russia" comes from the earlier, "Rus," the name by which the Vikings were known to the Byzantines and Muslims. (The origin of the word "Rus" is much disputed, as is so much else of early Russian history before the Tsars. Indeed, even Russia's Viking origins, as well as the influence of the later Mongol conquest, have been passionately disputed by Soviet and Russian historians as far back as 1749. Their protests, while understandable, are not very persuasive.)

One pernicious visage of the Viking overlordship that did remain, and which has had an influence comparable to the importation of Eastern Orthodox Christianity in the 10th century, was the essential separation of the leadership from the people. The Vikings were traders while their Slav subjects were farmers and woodsmen. This fundamentally separated the interests of the rulers and ruled. There is little evidence that this separation of interests has ever been closed, even into the late 20th century, although events since 1991 may be preview to change.

The Viking passion for trade and the natural route of the Dnieper to Constantinople strongly linked Kievan Russia with Byzantium. As the Viking identity blended with that of their Slavic subjects to beget the Rus, so too did the Rus blend with Byzantine influences to continue the transition to a distinctly Russian culture.

A century after Rurik's time his great grandson Vladimir was faced with the usual leadership problems of how to strengthen his country by improving the economy and protecting it from foreign threats. A strong alliance with an economically and militarily powerful neighbor made the most sense. There were a number of possibilities including the Muslims to the south, the Catholic Baltic states to the west, or the Orthodox Christian Byzantines to the southwest. In hindsight the choice seems inevitable, but the *Russian Primary Chronicle*, our principal written source for early Russian history, presents a picture of considerable deliberation, particularly in relation to the religious dimension of the issue, for Russia was a thoroughly pagan country at the time.

Vladimir questioned representatives of Islām, Judaism, Catholic Christianity, and Eastern Orthodox Christianity. From the Muslims he heard that God required the faithful, "...to practice circumcision, to eat no pork, to drink no wine, and, after death, promised them complete fulfillment of their carnal desires... Vladimir listened to them, for he was found of women and indulgence...But circumcision and abstinence from pork and wine were disagreeable to him. "Drinking," said he, " is the joy of the Russes. We cannot exist without that pleasure."" The Chronicle appears to rank drinking more important to the early Rus than an eternal heavenly satisfaction of carnal desire. That's a very strong drinking habit. Is it any wonder then that almost precisely 1,000 years later Communist Party General Secretaries Yuri Andropov and Mikhail Gorbachev successively found it impossible to get control of the national drinking problem.

Nor was Vladimir much impressed by the descriptions of the Catholic and Jewish faiths, but the Orthodox case did intrigue him. Nevertheless, he remained generally skeptical, so he sent emissaries to the countries practicing these faiths to observe matters first hand. As before, the Muslim, Jewish, and Catholic faiths failed to impress. On the other hand, the Byzantine Emperor, clearly recognizing the possibilities, pulled out all stops, offered his sister in marriage to Vladimir, and showed the envoys a spectacular Eastern Orthodox service. It worked perfectly. The envoys returned home awed, and they said so! Thus, according to the *Primary Chronicle*, Vladimir the leader of Kievan Russia was converted from paganism to Eastern Orthodox Christianity in 988 followed shortly thereafter by his nobles and many of the peasantry.

While some scholars consider the conversion story an apocryphal insertion, it makes good telling and very likely is true in essence if not in detail, that is, that Vladimir did very carefully consider the efficacy of conversion, and with equal care considered what to convert to. It is also interesting to note that the superficial elements of the Orthodox rite with all of its moving visual spectacle was a matter deemed impressive and potentially decisive, if not truly so in 988 then at least so by the *Chronicle* writers and copiers between then and the time of the oldest surviving version, the Laurentian text of 1377. As for the booze over sex matter, it's not entirely inconsistent with latter history.

Vladimir's baptism into Orthodox Christianity in 988 did far more than merely introduce Christianity into pagan Russia, it began a process of cultural transfusion that was to profoundly imprint and change its new host. For 500 years the Eastern Orthodox Church centered in the Byzantine capital of Constantinople had been evolving away from the Church of Rome. When encountered by the Rus it had already developed a belief in resigned suffering and acceptance of one's earthly condition. Patience and humility were seen as good, while it was bad form to want to better one's self, or to question one's betters. When the Slavs adopted these ideas and added them to their already compliant nature a new level of submissiveness to earthly power was reached. Add to that the submission to Viking leadership, the later Mongol domination, and finally the enserfment by the Tsars and enslavement by the communists and one has a very consistent 1,100 year history of a people thoroughly accustomed to be the object of absolute rule.

However, the Byzantine Church gave much more than that to Russia. It provided the seed for the characteristically Russian religious art and architecture. The round dome like that of the Hagia Sophia in Constantinople found its quintessential Russian expression in the onion shaped domes of her churches. The planar, full faced Byzantine religious painting became the icon of faith in every Russian church and home. And in addition to these outward forms, Russia inherited a certain messianic tradition from Byzantium as well. This was later expressed and expanded by a 16th-century monk who described the Russian Orthodox Church as the Third Rome carrying on after the Church of Rome and the Eastern Orthodox Church of Constantinople had become heretical and failed. Thus the Russians believed a special destiny devolved upon themselves to show the world the correct way to salvation. This was, in part, the same motivation that drove Lenin.

Also characteristically Russian, and the gift of the Orthodox Church, is the Cyrillic alphabet she uses. It was originally created in the ninth century by St. Cyril and his brother St. Methodius in order to translate the Bible and other religious texts into the Slavic language of the Bulgars, to whom they were missionaries. The Church simply extended its use to the Rus after their conversion, and since the Rus before this time had no written language, it de facto became such for them.

For 1,000 years Russia has struggled to find its place among nations. This struggle began with the Viking-Slav dichotomy in the ninth century and Russia's conversion to Eastern Orthodox Christianity at the end of the 10th. The net result by the early 11th century was that Russia had a distinct 'Eastern' non-European look that was to be accentuated later by two and a half centuries of Mongol domination and isolation from the west. (Figure 5)

We are now at the year 1000. Russia is becoming Christian, the city of Kiev is so dominant that we call the country Kievan Russia, and the Viking influence is largely submerged out of sight except for the artifact of the ruling class.

In the beginning it was a relatively good and prosperous time for Kievan Russia. The peasantry were largely free and the aristocracy preoccupied with their own interests in trade and the usual dynastic struggles to consolidate and expand their power. Contacts with Europe were far more common than they would be for the next 500 to 600 years as exemplified by the number of Russian princesses that became European queens including the queens of Norway, Hungary, and France, and from which, incidently, even today's Elizabeth II of England is descended. The city of Kiev was the equal or better of 11th-century Paris or London. Indeed, Kiev had a golden gate that dazzled visitors of the time and in the 19th century was musically immortalized by Modest Moussorgsky's *Pictures at an Exhibition - The Great Gate at Kiev*.

But problems were developing at home and abroad that spelled the end of Kievan Russia. Without some kind of institutional guarantees of freedom it's not surprising that the peasant would eventually be dominated by the naturally more powerful aristocracy, which is just what happened. By the 11th century the peasant's status had begun to decline to the point where he was known as a *smerd*, which at the time meant "stinker" and his apparent 'value' in law as measured by the fines for a peasant's murder was roughly equivalent to that for stealing a mare. Thus as the 11th and 12th centuries passed the free peasantry found itself being slowly enserfed. The land which they had owned was

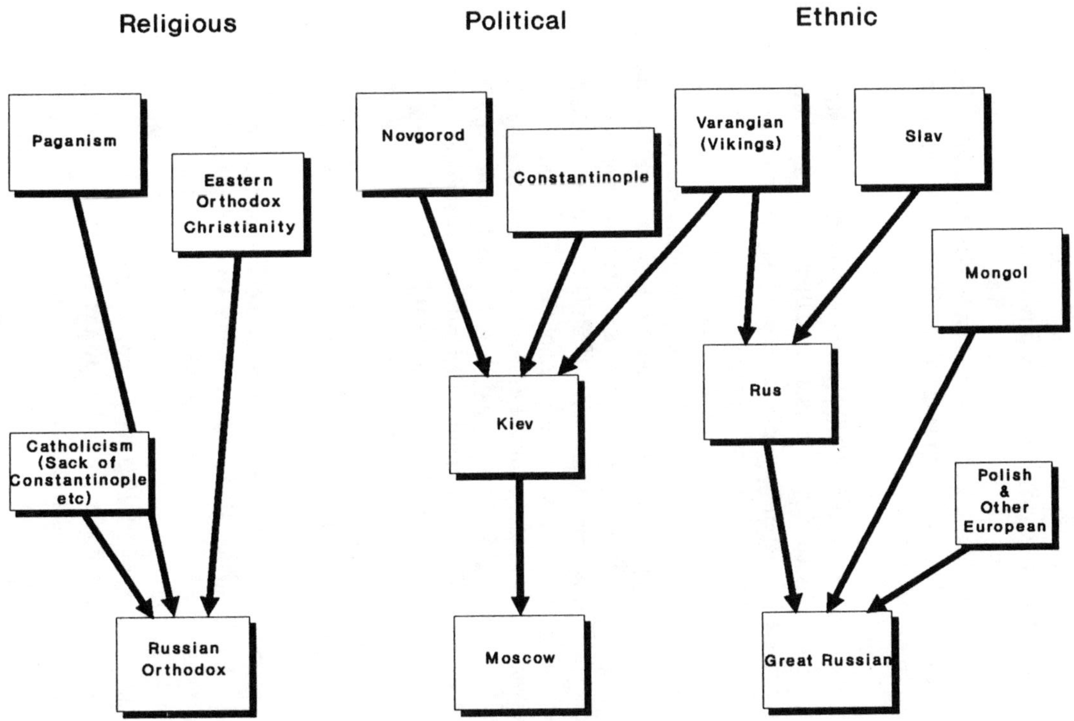

Figure 5
Influences on Early Russian History

gradually accumulated by the powerful as the peasant paid off his debts by surrendering the land and accepting various degrees of bondage. Large estate holdings became the norm for the aristocracy and the church.

At the same time peculiar systems developed regarding the inheritance of princely lands and the succession to princely thrones. They turned out to be especially disastrous. The first was called the appanage system. Appanage was a system where the ruler's principality was divided among his sons at his death, contrary to the custom in the West where the senior son got all or most of the fathers holdings. What this did was to break up originally large principalities into many smaller ones, each with its ambitious prince working to further his own ends. After a few generations any given principality would have been insignificant and easy prey for an outsider with political, economic, or military resources of any consequence. Of course conquest and intermarriage between princely families worked to consolidate separate principalities as well, but on balance by the early 13th century the Kievan state had fragmented into 12 major principalities.

The rota system merely added to the confusion and weakening of Kievan Russia. The rota system transferred the most senior throne, the Kievan throne for example, at the death of the eldest brother to next eldest brother who was an appanage ruler elsewhere. The next younger brother (if such existed) who ruled a still lesser principality then moved over to rule the second brother's principality, and so on in a kind of princely musical chairs of rule. Needless to say that this caused considerable conflict among brothers all of whom were experienced rulers, not to mention the dead-end disappointment of the sons of the deceased prince. The example of the chaotic rota system contrasts sharply with the concept of primogeniture succession customary in the West.

Other factors contributing to the decline and fall of Kievan Russia include a loss of trade with Constantinople after it was sacked in 1204 by Christians of the Fourth Crusade, a population decline caused by constant warfare between the appanage princes, attacks from surrounding countries and tribes, famine, and emigration - especially to the forests of the northeast. All this fatally weakened the state and set it prostrate in the path of a disaster unequalled in human history, the Mongol conquest.

In 1223 the Mongols appeared out of nowhere, utterly defeated the steppe tribes south of Kievan Russia, thoroughly trounced the Russian army at the Kalka River, and vanished into the eastern vastness of Eurasia as mysteriously as they came. (Map 5)

In the beginning the Mongols had approached the Russians asking for their neutrality and protesting no interest in Russia, only a wish to destroy the southern steppe tribes. The Russians made a colossal blunder, similar to that made by the Sultan of Khwārezmia, and executed the envoys. The Russians would later pay for this stupidity just as the Muslim world did for its arrogance, but for now they were simply bewildered by this sledgehammer blow out of the blue and naively thankful that the threat was gone.

The reality was not so reassuring. What the Mongols were really up to was a strategic reconnaissance of the lands north of the recently conquered eastern Muslim world. The raid was more or less incidental - until their envoys were executed. Then it became a very serious matter. In the meantime, however, other matters drew the Mongols back to their homeland northwest of China. But they hadn't forgotten how their emissaries had been handled. The cost to Russia would be incalculable and include 240 years of Mongol domination and thorough isolation from the West, millions of deaths, and the destruction of Kievan Russia forever. It would also be a major contributor to the rise of Moscow's political and cultural dominance, and the separation and independent evolution of Ukraine and Belarus.

In 1236 a Mongol army led by Batu, a grandson of Jenghis Khan, appeared west of the Urals again, this time with important business in mind, to permanently subdue the steppe and forest peoples of Russia before proceeding on to their primary objective, Europe. They erased the city of Bulgar from history and crossed the frozen Volga river in the winter of 1236/37. During 1237 they systematically attacked and destroyed the major northern cities of Moscow, Vladimir, Suzdal, and Ryazan among others. Ryazan was so thoroughly destroyed that when rebuilt it was rebuilt elsewhere. In the spring of 1238 the Mongol army headed for the northern jewel of Novgorod, but the spring thaw made the ground virtually impassible, and Novgorod was spared. Operations in Russia continued through 1239 and 1240, when outward bound for

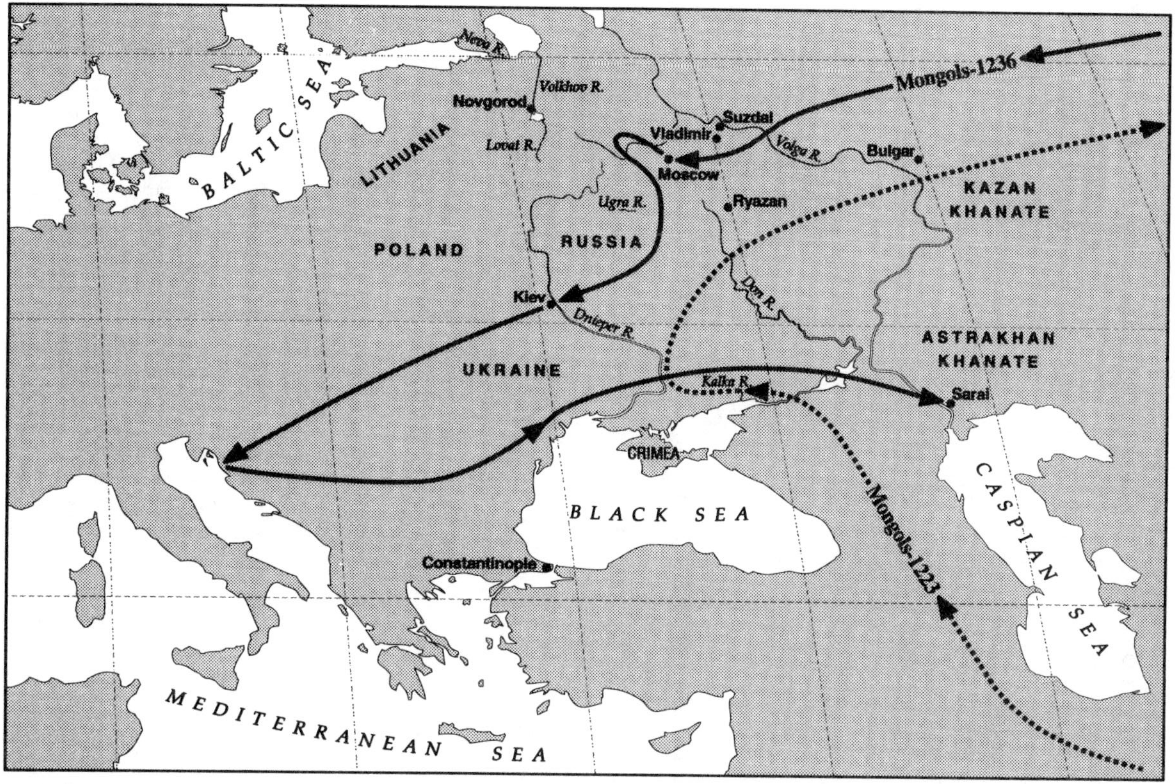

Map 5
The Mongol Invasions

Europe, they finished off the city of Kiev irrevocably ending whatever was left of any notion of a 'Kievan Russia.'

Kiev's fate was typical of many cities that resisted the Mongols. One Russian scholar estimated Kiev's pre-conquest population at 25 to 30,000, surely much too low. Another, more convincingly, estimated it at possibly well over 100,000. Six years later the Franciscan friar, John of Plano Carpini counted only 200 houses in the surviving town and described the surrounding area as containing, "...countless skulls and bones of dead men lying about on the ground." Accordingly, probably fewer than one in a thousand survived the Mongol visitation. The Mongol message this was intended to convey was that resistance was futile. If you surrendered you were usually spared. The cost thereafter was in taxes, booty, and slaves; a severe, but in no way unusual cost for the times.

Before the Mongols were finished with eastern Europe the Great Khan in central Asia died. Batu withdrew eastward beyond Russia to the region north of the Caspian sea where he established a 'capital' at the town of Sarai on the east bank of the lower Volga river. His army, known as the Golden Horde because of the color of his tent, settled in to enjoy the fruits of their conquest thereby beginning the era of the Tatar Yoke, traditionally set from the destruction of Kiev in 1240 to the 'ejection' of the Mongols in 1480. He didn't occupy Russia because it wasn't worth the trouble; better simply to tax and control the vassal princes under the Tatar Yoke. Russia wasn't on the most profitable trade route between East and West, nor were Russian forests as desirable an abode as the southern steppe for the Mongols. In fact, Russia wasn't even considered part of the empire by the Mongols, it was merely a place under its taxing authority, a vassal state not under its direct occupation.

Russia submitted, not without substantial 'encouragement' to do so, but nevertheless submitted, and without direct occupation. The Mongols provided the Tatar Yoke, but it was the vassal princes who applied it to their own people (and with the Orthodox Church urging submission). It was the princes who gathered the taxes for the Mongols, and it was the princes who vied with one another to curry favor with the Mongol chieftains at Sarai. The Mongols played the princes against one another in the classic divide and rule strategy. First they would favor one, then another, always with the intent to preclude one becoming too strong. Naturally among the more ambitious princes it became of paramount importance to flatter and cooperate with the Mongol leaders as much as possible. The princes of Moscow were clearly the most successful at this, and eventually received the Khan's *yarlik* (charter) to be grand prince.

With the disappearance of Kievan Russia into a dozen petty principalities each later seeking special favor with their Mongol overlords, and the destruction of Kiev itself, a leadership vacuum existed that Moscow was in a favorable position to fill. She had benefitted from the peasant fear of further Mongol raids in southern Russia with the subsequent peasant migration to the north in the region of Moscow. She had also benefitted from the effective rejection of the appanage system, and many of her princes were unusually long lived adding further to her political stability. Also her influence spread through a simultaneous slow but sure acquisition of additional territories by purchase, war, and inheritance. Over the period of the Tatar Yoke she had increased her lands from 500 square miles to 15,000 square miles. And perhaps most important of all, Moscow became the "Holy City of Russia" after the Orthodox Metropolitan moved his see there in 1328.

But other forces were acting on Russia during the 13th through 15th centuries as well. The historically most noteworthy was the constant warfare with the countries on her western borders including at least 35 'wars' with Germany, 44 with the Swedes, and 41 with Lithuania.

Lithuania eventually occupied large parts of western and southern Russia including much of old Kievan Russia. After intermarriage united the royal houses of Lithuania and Poland in 1386 these regions came under a period of Polish/Catholic influence that changed the culture sufficiently to permanently separate them from their old Russian heritage, thus the peoples we know today as Ukrainians and White Russians. Ukraine was not reabsorbed into Russia for almost 300 years and Belarus (as the White Russian nation is called today) was annexed about a century later in 1793.

It's not clear how long the Mongols would have continued to

dominate Russia if it were only a matter of the growing Russian strength under the leadership of the grand princes of Moscow. The historic trend was one of Mongol decline and Muscovite growth. On a few occasions over the next two centuries the Russians did meet and defeat elements of the Golden Horde, but these were minor defeats leading only to draconian retaliatory raids by the Mongols. The fact is that the Golden Horde had lost its *elan vital,* it was ageing and degenerating into petty interest groups. It lost its cohesion and weakened itself in family fights and became prey to Tamerlane's conquests in 1389 and 1395 as well as the ravages of the Black Plague. These processes all contributed to its continued weakening until the great moment of the non-battle of 1480 celebrated as the "Stand on the Ugra River." The uncrossable Ugra separated the Tatar and Muscovite armies who postured and futilely awaited reinforcements. Both armies finally gave up in frustration and withdrew just as the river froze and might have permitted a true battle. One suspects that neither army was so confident of victory that they wanted to test the issue. Nevertheless, this nonevent is often described as the moment of the final and decisive defeat of the Golden Horde. In fact, tribute, though of diminishing amounts, and only to selected sub khans, was continued for some time after 1480.

Numerous princes ruled Muscovy during the Tatar Yoke, but the most noteworthy was Ivan III (the Great) who took credit for finally breaking free of the Mongols in 1480. During his 43 year reign he very successfully continued his Muscovite predecessors' reintegration of the appanage principalities, and in 1493 took the title of "Sovereign of all Russia." He further assumed the title of "Autocrat" and began using the title of "Tsar" as he married the niece of the last Byzantine Emperor. He adopted the Byzantine double headed eagle as part of the Russian heraldry and began using elements of their court ceremonials. All this was in service of a conscious effort to extend his prestige and establish Russia's historical mission.

Indeed, with the fall of Constantinople to the Ottomans in 1453 the Byzantine Empire was no more, and its Orthodox religious see was discredited. This created an irresistible opportunity for Russia to claim the baton of 'true' Christianity. This was made explicit during the reign of his son Basil III when the influential monk Philotheus promulgated the idea of Moscow as the Third Rome. Russia was to be the third, and last, center of Christendom, to be the model and leader of the world to Christ and salvation after the heresy and conspicuous failures of Rome (Catholicism) and Constantinople (Eastern Orthodoxy). This notion very nicely fit the evolution of the Russian Orthodox church which only a few years before in 1449 had declared itself independent of the Patriarch of Constantinople and began appointing its own Metropolitan. Some sense of the importance of this move to independence, and especially its implications for the messianic role of Russia can be gauged by the fact that the atheistic Soviets officially celebrated the 500th anniversary of the event in 1949!

What were the consequences of the 240 years of the Tatar Yoke? Soviet historians have tried to minimize the impact of this episode of Russian history, but there is no avoiding the fact that it had a deep and lasting impact on all of Russian history to follow. The Mongol conquest not only applied the *coup de grâce* to the already decaying Kievan Russia, but provided many of the conditions which allowed Moscow with its clearly developing autocratic tradition to grow and become dominant. Without this situation it is easily conceivable that Novgorod and its relatively democratic tradition might have once again become the premier city of a new Russian state. Or perhaps the cities of Vladimir or Suzdal might have prevailed.

With the Mongols pushing the populations they didn't kill to the northeast, west, and northwest they helped, with Lithuania and Poland, to slice off and create the states of Ukraine and Belarus. At the same time they caused such a depopulation in the region just north of the Black Sea, and elsewhere in Russia that it would be centuries before the area would return to a population density comparable to that of Kievan Russia. Consider the forlorn picture presented by one 15th-century European traveler in Russia who described the land as woods and little hills empty as the desert with occasional signs of campfires left by other travelers, and, "...sometimes a little out of the way you shall find some small villages: but that is seldom." Catherine the Great was still concerned about the population vacuum in the Ukraine in the late 18th century.

Perhaps even more telling was the isolation from Western Europe forced on the Russians by the Mongols. Besides collecting taxes and slaves, and confirming princes in their rule, they made it a point of policy to forbid international contacts. This and the devastation due to the sack of their cities and the decimation of the population largely arrested Russian cultural development for two centuries. Ever since, Russian "backwardness" has been proverbial, and as we shall see, there has been a long series of 'catch-up' episodes which in the long run have been largely unsuccessful.

The Tatar Yoke was also a 240 year lesson in humility and national enserfment. The Russian peasant learned through 10 generations that survival was achieved through patience and suffering, and that relief came only in a hereafter offered by the Church. The Russian viewed himself in one of the old Chronicles as descended, "... from the gentle and humble Jacob, and therefore we cannot oppose them (the Tatars, descendants of Ismael), but humiliate ourselves before them, as Jacob did before Esau. Yet we defeat them with the arms of the Cross: this is our help in battles, and our support against our enemies." In order to better control the country the Mongols astutely facilitated the growth of the Church by giving it special privileges including exemption from taxes. With the acquisition of land and influence the Orthodox Church became the natural ally of the grand prince while encouraging the peasant to accept his lot and offered in compensation notions of a national messianic mission. By the reign of Ivan IV beginning in 1547 the conquest and centuries of domination by the Mongols were explained in an official pronouncement as, "...wilful humiliations designed to spare the people and expiate their sins." Elsewhere the Russians were cited as, "...famous for their laziness, so that a single Tatar could capture many Russians." While this is clearly an exaggeration it does suggest a certain compliant quality.

In the next chapter we shall find with the emergence of the Tsars that peasant suffering didn't end with the expulsion of the Mongols. We shall also see the growth and continuation of themes in Russian history that will later become particularly familiar in the 20th century, oppressive state police, leadership paranoia, and social chaos.

Chapter 7

Tsarist Russia: the Beginning 1547 to 1696

The grandson of Ivan III was a very troubled child who characteristically and portentously amused himself by torturing his pet animals. Whether this behavior sprang from organic causes or from the neglect and danger he experienced as a youth we cannot say. His father died when he was three, and his mother was assassinated when he was eight. He grew up in a world devoid of personal feelings where political power meant everything. As the crowned grand prince under the guardianship and regency of his mother, and then after her murder, under the avaricious and power hungry boyars, he couldn't possibly have developed much trust in those around him nor much value their lives. And this was to be so with all those that he ruled. What we can say is that in 1547, at the age of 16 he decided that enough was enough and had himself formally crowned Ivan IV, Tsar and Autocrat of all the Russias. He was the first to be formally crowned "Tsar," a word derived from "Caesar" but with a Russian meaning closer to "Emperor." His contemporaries would later add the sobriquet, *grozny* which specialists often remind us means something like 'awesome' as in a characteristic of majesty. Nevertheless, a looser translation of *grozny* is the one history has firmly and deservedly attached to his name; thus we know him as Ivan the Terrible.

The legacy that earned him this title is long and varied. His reign began in a relatively mild way no doubt tempered by the still strong boyars, but for Ivan the Terrible 'relatively mild' still meant hundreds of executions as well as having the tongue cut out of one nobleman who offended him by "uttering rude words."

While Ivan's grandfather, Ivan III, had largely released Russia from the Tatar Yoke, the Tatars (Mongols) continued to be irksome neighbors. By the mid 16th century the Golden Horde had fractured into several sub-hordes and become so weakened that the conquest of the Volga River region north of the Caspian Sea became feasible. Accordingly, in 1552 Ivan conducted a successful campaign against the Kazan Khanate, and four years later with equal success against the Astrakhan Khanate. The Volga River and his eastern border were now more or less secure. (To celebrate these victories he built one of the marvels of Russian architecture, the church of St. Basil in Red Square.) Such, however, was not the case elsewhere, for by 1559 his campaign to the south against the Crimean Khanate failed. Indeed, the Crimean Tatars remained strong enough to attack and briefly occupy Moscow 12 years later in 1571. Their success, however, was as much a measure of the chaos and growing catastrophe of Ivan's brutish rule as of Tatar military prowess.

The failure of the Crimean campaign was not Ivan's worst military adventure, however, it was the exhausting 24 year Livonian War. The eternal Russian quest for a year-round ice-free port led Ivan to invade Livonia, the region now encompassed by parts of Estonia and Latvia. If successful he would have the ports of Narva and Riga for direct trade with western Europe. While this would not get him to a true "blue water" port (since the Gulf of Finland was still ice-clogged several weeks of the winter) it was an infinitely better port of entry than remote Archangel on the Arctic Ocean's White Sea. His invasion began in 1558 with such success that Narva was quickly taken. Subsequent European trade increased so rapidly that spectacular price declines of goods imported into Russia soon followed. But by the 1570s Polish, Lithuanian, and Swedish resistance had generated a virtual stalemate, and later counter attacks forced Ivan to accept a settlement in 1582-3 that cost him everything he had gained in the beginning.

Among Ivan's other notable actions during his 37 year reign was the creation of the *oprichnina*. *Oprichnina* is the word for his personal holdings within Russia which he acquired by confiscating estates of the nobility and the Church. The *oprichnina* eventually amounted to almost half the country's land. *Oprichnina* or *oprichniki* also applies to the personal bodyguard and army of 6,000 picked men characterized by their ruthless devotion to his whims and their gratuitously cruel oppression of the people. In following Ivan's orders and their own savage instincts the *oprichniki* remind one very much of Hitler's SS and Gestapo, even to their black uniforms and the use of a dog's head insignia like the SS's death's-head. Terror was spread across the land unrestrained by any law, but perhaps their most profound impact came when Ivan, fearing treason among the historically independent minded people of Novgorod, used them to bring about the final destruction of Novgorod's political, social, and commercial infrastructure. Tens of thousands were slaughtered in the city which had been Russia's main window to the West. Thereafter it was permanently shuttered closed resulting in further cultural isolation that only added to Russia's historic difficulty in connecting with the West, notwithstanding the new trade route he had allowed to develop earlier through Archangel.

In 1572, after seven years of growing power and terror, Ivan destroyed the *oprichniki* itself. It's not clear whether the motive was an appreciation of their debilitating impact on the country that included the permanent destruction of many noble families and the flight of so many peasants that depopulation in central Russian became a problem, or his paranoia that they might turn on him.

Ivan's religious passions never thoroughly tamed him even though it is said that he spent up to five hours a day, every day, in church. Indeed, his reign marked the beginning of the end of an independent Orthodox Church (if one could say that the Orthodox Church in Russia was ever independent of the ruler.) Control, or at least neutralization of the Church was increased by his confiscation of considerable Church land and his wringing from it promises not to interfere with his persecution of those he believed to be traitors, no matter how flimsy the evidence. The metropolitans gave in to this demand so easily that one suspects they were more interested in maintaining the ruler's favor than in providing a moral beacon for Mother Russia.

Not only did the process of enserfment of the peasants continue under Ivan IV, but virtual enserfment of the nobility developed as well. Unlike the nobility elsewhere in Europe who were largely independent, Ivan instituted compulsory state service for his nobility, and at their own expense. The wife of one prince couldn't even appear at the Tsar's court as commanded because her husband had pawned all her clothes to finance his military service. This substantial financial burden naturally fell eventually on the peasantry. When the peasant was forced to borrow money to pay taxes the land owners (the nobility and the Church) loaned it. Not surprisingly, the peasant often wasn't able to repay the loan and thus became permanently indebted to the land owner. This peasant indebtedness was one of the most important instruments of enserfment, and it provided a nice guarantee of labor, especially on those lands that were becoming depopulated by peasant flight in the face of Ivan's misrule. In fact, this technique of binding the peasant to the land worked so well that in 1586 they were officially forbidden to pay off their debts. Like the popular American song of the 1950s lamenting the plight of coal miners, 'they owed their soul to the company store.'

All in all the reign of Ivan IV, the Terrible, was a disaster for Russia. The people, peasant and noble alike, were brutalized, impoverished, and enserfed. Though the expansion to the east that would eventually extend Russia all the way to the Pacific was beginning under his rule, it happened more or less spontaneously and not the result of his conscious policy. Ivan began the final reduction of the Church that undermined its moral authority, he accelerated the momentum of rule toward absolute power beyond anything known in Europe, he aided in Russia's isolation from the West through the destruction of Novgorod thus continuing into the future the habitual backwardness of the land, he furthered the conceit of Russia's special role in history by contriving a silly genealogy, the *Book of Degrees* that was saturated with nonsense ranging from claims of the Tsar's lineage going back to Rome's Augustus Caesar to a miraculous birth for Ivan, and he worked to reenforce the notion that suffering was somehow a salubrious experience. Contrary to the tendencies of many Russian historians who judge Ivan IV's reign as not all that bad and perhaps comparable to his near contemporary, England's Henry VIII , one finds it difficult to escape better comparisons to Hitler or Stalin, the differing historical eras notwithstanding.

If Russian suffering was trying under the Tatar Yoke and difficult under Ivan the Terrible, it was severe during the "Time of Troubles." The Time of Troubles was arguably another gift of Ivan to his country. It originated in part out of the power vacuum that followed in the decade after Ivan's death in 1584. The problem was, he left no viable heir to the throne. He did have a viable heir until 1581. His second son, Ivan Ivanovich (his first son, Dmitry, drowned as an infant) showed potential for some day becoming a good ruler. Unfortunately for Russia, Ivan in an argument over how his son's wife was dressed, struck the 27 year old tsarevich in a momentary fit of rage and killed him. After Ivan's death there followed a 14 year non-rule of the incompetent Tsar Feodor I, the youngest son of Ivan IV. With Feodor's death the line of Russian rulers tracing back to Rurik ended, and the boyar regent, Boris Godunov, the real ruler behind Feodor, emerged and claimed the throne.

The Time of Troubles began in 1598 with Boris Godunov's "election" as Tsar by the zemsky sobor. The zemsky sobor was an assembly of boyars and Church officials appointed by the Tsar on an ad hoc basis. The tradition of calling such gatherings probably went back 150 years to Ivan III's time. Some historians like to point to the existence of such councils as indicative of the limitations of the Tsar's power, or even as some tincture of democratic rule in Russian history. (A similar and only slightly more persuasive assertion might be made regarding the duma, a sometime legislative and administrative council of the nobility.) But the zemsky sobor was nothing of the kind. They were appointed by the Tsar and sat at his pleasure alone, they had little or no institutional tradition, and no well defined procedures, role, or authority (again sharing many weaknesses of the duma). They were mostly used by the Tsar to force the appearance of consensus in support of his actions and rarely acted otherwise. When there was a power vacuum, as in the Time of Troubles, they did act independently, but it was a matter of action by default since there was no Tsar on the throne when they elected Godunov, nor was their a reigning Tsar in 1613 at the end of the Time of Troubles when they elected the first Romanov Tsar. It is also true that the zemsky sobor largely ran Russia during the 17th century up to the time of Peter the Great, but this was once again a matter of influence by default since the early Romanovs were unusually benign and passive rulers.

The Time of Troubles lasted 15 years from 1598 to 1613. It was a time of anarchy, famine, pestilence, banditry, civil war, and foreign occupation, often all at the same time.

When Boris Godunov took the throne he continued Ivan's struggle to weaken the nobility and strengthen the power and authority of the Tsar. By the time the Romanovs ascended the throne true absolute power was essentially at hand, if the Tsar wished to use it. He further restricted peasant rights while loading them with still higher taxes as well as the additional burden of more military conscription. He continued to develop trade ties with England and Western Europe and recognized that Russia was backward and needed an infusion of Western ideas. He sent a number of students abroad to study, however, he shared a problem with the 20th-century Soviet government in that the students were disinclined to return home afterward.

Tsar Boris' bad luck went far beyond the loss of students sent abroad, however. The crop failed for three years in a row causing a famine that may have killed one third of the population. And right in the middle of this an epidemic, probably typhus, occurred in 1602 killing hundreds of thousands more. Significant numbers of those who survived were driven into brigandage. As his measures to control this lawlessness were felt, many from peasant to prince began to express the already existing feeling that he had no right to the throne. At this moment a person calling himself the Tsarevich Dmitri invaded Russia.

In 1604 Poland produced a pretender to the throne, the "False Dmitri." He was impersonating Ivan IV's questionably legitimate, deceased youngest son, also named "Dmitri" like Ivan's first son. (His legitimacy was at issue because he was the offspring of the union of Ivan and his seventh(!) wife, a union questioned by the Church.) In 1591 this true Dmitri died under suspicious circumstances, some believing that the ambitious regent, Boris Godunov, was involved. Whatever the truth, Moussorgsky's wonderful opera, "Boris Godunov" has pretty much sealed poor Boris' guilt in the popular mind.

Godunov probably wouldn't have had much difficulty defeating the False Dmitri, but with more bad luck Godunov suddenly died. (History is vague about how. His health hadn't been good for a few years when he apparently had some kind of stroke getting up from the dinner table. Rumors have said suicide, poison by another hand, and natural causes - take your pick.) Dmitri became Tsar almost by default. The next eight years were a mess. False Dmitri #1 was killed, and the boyars enthroned one of their own, Prince Basil Shuisky. Meanwhile, the Cossacks were producing another False Dmitri (#2) who the ever helpful Poles also supported. Shuisky called in the Swedes for help (*déjà vu* Rurik 750 years before), but that didn't save Shuisky from being deposed by his own people in 1610, nor save the life of Dmitri #2; nor stop Sweden from taking some choice real estate from Russia on the Gulf of Finland. Nor could Poland resist the opportunity of such chaos; she invaded and occupied Moscow, and that's how things sat for three years (1610 - 1613); no Tsar and no capital. Meanwhile, virtually every interest group in the country was fighting for power, "Time of Troubles" indeed.

What Russia needed was some institution or idea that would unify the competing factions at least long enough to get the Poles out of Moscow. It turned out that the Orthodox Church, religion, and nationalism were the institutions and idea to do the job. The Poles were doubly disliked at that moment, not only because they were occupying Moscow, but because they were Catholic. For a Russian Orthodox Christian there was nothing worse than a Catholic except possibly a Polish Catholic. Accordingly, the Russian Orthodox Church provided the leadership and inspiration to arouse and unite the people against their common enemy, and after one abortive effort a Russian army finally retook Moscow in the autumn of 1612. The next order of business was to get a government going.

The zemsky sobor met in 1613 and chose Michael Romanov to be Tsar. This 16 year old member of a prominent boyar family was selected partly in the hope that his youth would make him a compliant servant of the boyar class. They got their wish and then some for Michael and his three immediate successors were weak rulers; it would be 83 years before a Romanov, Peter the Great, would totally reassert the full power of the Tsar. The 300 year Romanov dynasty, started by Michael and ending with the 1918 murders of Nicholas II and his family, would see only one other substantial monarch, Catherine II.

Russian history continued to be a cacophony of war, faction, and peasant suffering, but a stable monarchy did provide the foundation on which slow but sure national growth developed. The three most important events of the century before Peter the Great were the expansion east to the Pacific, the recovery of eastern Ukraine, and the schism of the Russian Orthodox Church.

Russia first became a significant power at the time of Peter the Great. An important cause of this, or at least an important sustaining cause, was its enormous size and the resources that went with it. A century earlier under Ivan IV Russia was already larger by far than any European state. The later expansion east more than doubled its size making it the largest nation on earth, as it remains to this day. This movement east began very gently in the fourteenth and fifteenth centuries with the growth of monasteries on the north and northeast frontier, partly driven by the Tatar Yoke. It was accelerated in the 16th century with the flight of peasants from Ivan the Terrible's persecutions. It was completed in the 17th century by Russian adventurers and peasants seeking furs, empire, freedom, and land.

The door was opened when Ivan IV conquered the Tatars of Kazan and Astrakhan. This made the frontier for trans-Ural expansion safe. After the Time of Troubles Russian exploration and settlement in Siberia developed steadily. Beyond the Urals the land was peopled, when peopled at all, by relatively primitive tribes who were easily subdued. The Pacific Ocean was first reached in 1639 and the Amur River in 1644 where the Chinese were encountered; the first Russian sailed the Bering Strait in 1648, and the last corner, the Kamchatka Peninsula, explored in 1697. Siberia was now in hand to provide Russia with trade connections to south and south west Asia through Samarkand and Bokhara as well as with China and all of far eastern Asia. Trade and furs were the early rewards of Russia's Siberian possessions and later, immense mineral wealth.

While Russia was expanding east it was also recovering lands to the west and south lost to Lithuania and Poland during the Tatar Yoke. The Ukrainians were not at all comfortable with Polish attempts to integrate them into Polish culture, even after 300 years of Lithuanian and Polish control. They were particularly incensed by Polish attempts to convert them to Catholicism or to force them to accept the Uniate Church, a kind of Orthodox Church with significant links to the Church of Rome. Several rebellions developed and Ukraine appealed to Russia for help. The Tsar was rather luke warm to the idea, not relishing another confrontation with Poland, but eventually he came around under the prodding of the zemsky sobor. In 1654 Ukraine's request for union with Russia was accepted, and after the usual maneuvering and battles, the Poles reconsidered the burden of hanging onto an area so disinclined to be held, and signed over eastern Ukraine to Russia in the Treaty of Andrusovo in 1667.

Alas, Ukraine had been away from the home stock for too long. The Ukrainian mother of Russia and her historic Kiev, the "mother of cities," were no longer Russian. Ukraine had developed a national identity of its own during its 300 year absence. Though this would not be an immediate problem it certainly became one in the long run. For example, Ukraine has always claimed that she joined Russia with the understanding that she remain autonomous; Russia has always maintained that Ukraine was not autonomous and Ukrainian autonomy was not an acceptable condition for the Tsar's "protection." However, in the mid 17th century this was a mere detail as Ukraine finally got out from under the oppressive Catholic Poles. In the meantime Ukraine improved Russia's geopolitical position in the face of the ever present and irritating Ottomans around the Black Sea, and added substantial material wealth to Russia with its agricultural abundance.

But the greatest long term impact on the people themselves, and their relationship to the government, resulted from a conflict within the Russian Orthodox Church that culminated in the Great Schism in 1667.

An ambitious new Patriarch of the Russian Orthodox Church had been looking to increase his authority with the Tsar in part by facilitating Ukraine's return to the Russian fold. This could be done by making their respective Orthodox Churches as similar as possible, but in three centuries their Churches had diverged slightly. Religion was a serious matter in Russia, more serious than even the new Patriarch imagined, for when he tried to make some slight changes in the Russian Orthodox texts to coincide with what he believed were more accurate Ukrainian versions he stirred up massive resistance. It started in 1652 with textual changes and continued to 1656 with slight changes in ritual. Patriarch Nikon was accused of heresy in 1663 and relieved of his post by a Church Council in 1667. However, this same Council approved of the changes. This was the last straw for many in the Church who believed that Russia with her isolation from the corrupting West was a better protector of the true religion, text, ritual, and all, than any Greek revisionists sneaking in through the Ukrainian Church. The Church fractured into two factions; the Old Believers who resisted any change whatsoever, and the remaining majority Russian Orthodox Christians who accepted the changes. The Church reform was supported by virtually the entire senior clergy as well as by the Tsar and his government. Though some have remarked that this represented a curious reversal of the situation in the Reformation where it was the people who wanted reform and the clergy who were conservative, such was not truly the case. In Russia the Great Schism was essentially a difference of opinion in degrees of extreme conservatism. The so called reform which so upset the Old Believers amounted to matters such as the spelling of the name of Jesus, whether two or three fingers should be used in making the sign of the cross, the number of times "Hallelujah" should be said in the Liturgy, and the direction of processional marches. It is very difficult to understand how matters so trivial could divide so deeply, indeed divide and impassion enough to cause three centuries of oppression and killing to the degree that in the beginning tens of thousands of Old Believers chose to immolate themselves by burning to death rather than accept the changes. Clearly, something very important is being revealed by this episode of Russian history.

It is often said that Russia and its history are saturated with religion, and that it is impossible to understand Russia without understanding the meaning of religion for her people. Though this is undoubtedly true, I think there may be an even more profound issue at hand here, the fundamental Slavic character. It is out of this basic character that their particular needs and manifestation of Orthodox religion make best sense.

Contemporaneous and later descriptions of the early Rus Slavs included words such as meek, gentle, long suffering, and humble. No wonder then of the appeal of a self sacrificing Christ when Christianity was presented to them. Also, one cannot help but recall that the early, pre-Rus Slavs were so frequently dominated and bound to the service of others that the word "slave" was derived from their name. Interestingly, they also had no god of war. Then historic Kievan Rus was formed, not spontaneously by the Slavs, but by the energy of foreigners, the Vikings. For a century thereafter a land of pagan peoples began to cohere. As political matters dictated, the Kievan prince Vladimir examined several religions for the most useful for the needs of his rule. Eastern Orthodox Christianity was the choice, but it came to early Russia not through direct missionary penetration of the peasantry, but by the conscious selection of the rulers. It was brought in by the State, for the State, and later, of the State. Not surprisingly, the Russian Orthodox Church rarely deviated from State interests. Over the centuries the power of Church and State would seesaw back and forth, but the Church's predominance over the State almost always occurred by default of secular rule. In the interim, it liked to use the Byzantine concept of cooperation, a "symphony of powers," to forward its authority.

Descriptions of the 'conversion' of the Kievan Rus usually don't give a clear picture of what really happened. Some write of the rapidity of conversion and how it was freely accepted. Others remark on the centuries long effort and significant coercion required. These differing descriptions can be explained by noting that the leadership converted virtually immediately, and converted largely for reasons of state, while the peasantry, not caring about reasons of state, was strongly inclined to retain its pagan beliefs. When the peasantry was put under pressure to convert it often did so as the peasantry frequently do everywhere, it simply combined the new religion with its pagan beliefs. This happens with every religion. The persistence of substantial pagan belief among the peasantry continued into the 14th and 15th centuries, and was reflected in the persecution of the sect called the *Strigolniki* some of whom were put to death in 1375 in Novgorod for heresy. Some of their beliefs continued even to the time of the Great Schism and beyond, for example in the practice known as confession to the Earth. In later times confession to the Earth (a substitute for the sacrament of penance) was rationalized as acceptable in the absence of clergy, for example when in remote Siberia, or with the Old Believers who also were largely without clergy. The net result of all this was that superficial conversion took place relatively rapidly, but that deep in the Russian character conversion was highly selective and always in service of those qualities preexisting far back into the shadows of their pagan past. The Slavs were a much put upon people, accustomed to suffering and submitting. They seemed to have reached a psychological accommodation with this state of affairs well before Christianity arrived, and indeed, selected a form of Christianity, Eastern Orthodox Christianity, that was supportive of this accommodation. When the official, institutional religion failed them, the peasantry simply used the spectacle of the Liturgy for their inner needs or ignored the institution altogether and continued their practice at home in front of the icon corner. Accordingly, when one considers Russia and its religion, one should never confuse the historic institution with popular belief and practice. Each has had its role, and they often were quite different.

Another example of this difference occurred in the late 1400s when some monks led by Nil Sorsky declared that it was wrong for the Church to be so intimately connected with the State and to hold land and serfs. They were called Non-Possessors. Their opponents, the Possessors, believed that it was not only acceptable to have all this wealth and power but that it was essential for their religious function. Ivan III supported the Possessors, especially since they believed so strongly in the mutual support of Church and State. A Church Council officially settled the matter in favor of the Possessors in 1503, but the matter never really died, and was part of the under current in the Great Schism where the question of the integrity of the Church was never far below the surface. The Non-Possessors naturally became the Old Believers.

One very important case where the beliefs of institutional and personal religion coincided was in the messianic role of Russia and the Russian Orthodox Church. With the Third Rome concept it became easy to believe that the only true and pure religion left in the world was that which was practiced in Russia. Add to this Russia's isolation that protected her conceit, and no doubt a good dose of compensatory self-aggrandizement in the face of several centuries of foreign subjugation, and one gets a virtually irresistible notion of self importance. (This theme is not unique to the Russian experience, however, recall the views of the Muslim world as well as those of the West.) We shall see this theme much more explicitly when we consider 19th-century Russia.

The Great Schism was caused because the State wished to be

flexible in order to acquire eastern Ukraine. A sizable part of the populace, including the Old Believers, wanted to keep the ritual unchanged and in the practice of their fathers and ancestors. When the official Church insisted under pressure of compelling reasons of State, the peasantry did the usual thing and discounted the Church. The result was that Church influence with the peasantry was considerably weakened, and it became ripe for its final subjugation to the State. In a generation, under Peter the Great, the Tsar would finally be undisputed and absolute ruler of Russia.

The time of the first Tsars, Ivan III to Feodor III (16th & 17th centuries), was an era of growing independence from other powers, of developing national identity and geographic expansion, of peasant enserfment and the semienserfment of the nobility, of political co-option of the Russian Orthodox Church, of continued internal and external conflict, and of substantial lag behind the West in cultural development notwithstanding several attempts to catch up. In the first half the land was significantly depopulated as the Four Horsemen of the Apocalypse; war, famine, pestilence, and death rode at will. In the second half a slow recovery began under the rule of a series of relatively benign Tsars. With these few generations to recover, Russia was now ready for another energetic Tsar.

Chapter 8

Tsarist Russia: the Beginnings of Empire 1696 to 1800

Seventeenth-century Russia was a primitive and mysterious land. It was a place to which adventurous western voyagers and merchants traveled and wrote about as though describing darkest Africa. It was a land still gripped in ways of life reminiscent of medieval Europe 300 years earlier. The cultural transformations of the Renaissance and the Reformation had not touched Russia. It was a land of primitive, unrealized power manifest in the steady and irresistible outward pressure against its frontiers, its ferocious animal like squabbling over superficial religious detail, and its dark and brooding Tsars and nobility. The nobility lived only to serve the Tsar and were obsequious well beyond the custom found in Europe. As one English traveler put it in 1698, they had "...infused into them the principle of passive obedience, in so transcendent a degree, as not to have the true sense of that liberty we enjoy in these parts." It was a land so dominated by severe continental winters that from earliest times the dress of choice was fur. The Tsar's court was often described in terms of attending boyars embedded in floor length coats, tall pointed hats, and massive beards. Women of nobility were cloistered as if in Muslim harems. The Orthodox Church enticed if not commanded allegiance with its glorious but endless liturgies. Tsars were reported to spend several hours a day in church, though their presence was not always in religious devotion; affairs of State were conducted there as well. Such, in part, was the Russia into which Peter I was born in 1672.

Peter was the only child of Tsar Alexis' second wife. As such he was largely left to amuse himself while his half brother, Fyodor

III, ruled after their father's death in 1676. Then Fyodor died when Peter was 10. When the family of Peter's mother tried to make Peter the sole Tsar in spite of the existence of another, older half brother, Ivan V, a coalition of forces powered by the *Streltsy* palace guards revolted and forced a co-reign with Ivan as senior co-Tsar and Peter as junior. Since at that moment both were children, Peter's half sister, Sophia, was named regent.

The revolt to which the 10 year old Peter was witness was exceedingly bloody for Peter's supporters. The *Streltsy* behaved as a savage, dumb mob only partially sensing its own conservative notions about who was the legitimate Tsar. The *Streltsy*, or musketeers, had originally been professional military units organized from the peasantry many years after the disastrous *Oprichniki* guards had been disbanded by Ivan The Terrible. Now they represented the most reactionary elements in society protesting their unconditional support for the Tsar, yet expecting, if not demanding, a certain behavior from him as well. The Russian peasant, and accordingly the *Streltsy*, thought of the Tsar rather as an icon, an image of God to be venerated, cherished, and obeyed, but also to be possessed, for the Russian peasant possessed God as much as God possessed the peasant.

With Sophie as regent and Ivan as senior Tsar, Peter continued to be left alone, albeit the situation necessarily had its tensions. So who was this person who would be, arguably, the most influential ruler of Russia in its 1,000 year history?

The adult Peter was tall. At six foot seven inches he towered over virtually every person he ever met. He was intelligent, but the freedom given him in his youth allowed him to escape much formal schooling, an escape he frequently regretted later. He was immensely inquisitive about practical matters and technology. All through his life he would take the trouble to stop what he was doing to ask about the how and what of whatever chanced his way. Often he went far out of his way to see and learn of new things. He was unpretentious and liked to work with his hands, especially among artisans and shipwrights. Rank and social standing meant little to him, he judged a person solely on their merit, how well they did their job. He was often generous to others but privately his tastes tended to the austere. He valued and rewarded personal and professional honesty, and detested the corruption endemic in his government even while he had to accept it. He was independent, his own man, and often disregarded the conventions of his time that were continually pressuring him to be a conventional Tsar. Perhaps more than anything except his unusual height, he was a man of tremendous energy. He did everything with great energy, whether it was traveling, or learning a skill, or building a navy or a city. On the less flattering or darker side he was sometimes unusually cruel. The subjects of this included even his son and heir apparent who died from judicial torture administered to reveal a fancied plot against the throne. And he was prone to personal violence with those around him (but usually quick to forgive.) This did not, however, extend to his second wife, Catherine, whom he treasured.

About the time of his 16th birthday Peter discovered an old, decaying English boat in storage on one of the Tsar's estates. After having it restored he learned to sail and quickly developed a passion for boats and sailing that continued throughout his life. This passion also led to his eventual creation of the Russian navy. Years later Peter preserved the old English boat nostalgically calling it the "grandfather of the Russian navy." It still exists and is on display in St. Petersburg today.

In 1689 the fickle *Streltsy* once more acted, this time in deposing the regent Sophie and turning the operation of the government over to the partisans of Peter. Ivan and Peter continued as co-tsars but were only symbolically involved. Both were quite happy with the arrangement for it allowed the sickly and simple Ivan a comfortable and undemanding private life and the energetic and robust Peter full opportunity to pursue his passions for carousing with his foreign friends in the German Suburb and for sailing. For the next five years these pastimes would mold Peter's view of the world and lead directly to a shift in the Russian tide of history as Peter hurled his motherland towards the West.

Peter took active control of the government in 1694 when his mother died, though technically he remained co-tsar with his half brother. Once again the situation suited them both, and Peter never considered deposing Ivan whom he always regarded with great and tender affection.

Very quickly the young Tsar decided that his first military

venture would be to the south where the Crimean Tatars, supported by the Turks, continued to harass his southern frontier. Perhaps an even more powerful attraction for this Tsar was the Black Sea. At the time the Black Sea was essentially a Turkish lake completely surrounded by Ottoman lands. If Peter could punch through an access down the Don River and beyond the Sea of Azov he might have the personal pleasure of sailing a large ice-free sea and the political pleasure of challenging a decaying Ottoman Empire for territory and access to the world beyond. After an initial failure he did take the town of Azov on the Don near its mouth on the Sea of Azov. Eventually a peace was concluded ratifying Peter's gains (albeit access to the Black Sea itself would not come in Peter's lifetime). This victory against a world power, minor though it was, not only hugely gratified the Russian people who hadn't beaten anyone in some time, but it also amazed Europe.

It had been only a few years since the last Ottoman threat to Vienna in 1683, so Europe was pleased to find a new ally in Russia. And Europe was not only pleased but surprised, for up to this time Russia had been more mystery than substance. The feeling was reciprocated as Peter also savored the possibilities of European alliances against the Ottomans.

With the natural death of Ivan in 1696 Peter became sole ruler of Russia and in complete control of affairs. The several currents of his young life were coming together to carry him and Russia irrevocably onto the world's stage; his interests in sailing, in the sea, and in building a navy, his ambitions to the south against the Crimean Tatars and the Ottomans, his recognition of the vast superiority of Western society and technology (through his contacts in Moscow's German Suburb), and his recognition that Europe could be decisive in realizing all of these for Russia inspired Peter to organize a Great Embassy to Europe in which he would personally take part. The Great Embassy was one of the earliest of a long line of 'catch up' efforts undertaken by Russia through her history.

Peter I was the first tsar ever to travel outside of Russia. The Great Embassy was organized to seek allies against the Ottomans and to find and bring back as much technology, especially naval technology, as possible. Since Russians were largely ignorant of ship building he not only commanded that several dozen Russians go with him to learn the art, but also hired as many as 1,000 European artisans expert in these and other skills to go back to Russia. In order to avoid the pomp, ceremony, and distractions of royal life Peter himself traveled with the Embassy officially incognito, though it was obviously not practically possible given his distinctive height. Nevertheless, he did manage to work with his own hands in Dutch shipyards and otherwise indulge himself in much common labor, all the while learning the essentials of several important trades for his country. In this he was very successful, but insofar as establishing an alliance against the Ottomans he failed. After a year and a half of travel in the West, even as far as England, Peter returned to Russia abruptly in 1698 upon news of another *Streltsy* uprising.

Enough was enough, the *Streltsy* had lost its military prowess and remained only a constant irritant and threat to the Tsar. Peter returned, found the leaders of the latest rebellion, had them mercilessly tortured to reveal its complete extent, then had a thousand executed, many with extreme cruelty. Perhaps remembering the dangers of Ivan IV's *Oprichniki* he then disbanded the *Streltsy* thus relieving Russia of the 'Praetorian Guard' syndrome for a generation. (Unfortunately, as we shall see shortly, it didn't take long before a new military power center evolved in the Imperial Guards and became instrumental in deciding Russia's leader three times in the next 100 years.)

Peter's return began the great westernizing epic for Russia and set in motion internal strains between Western and Russian ways that have not yet been completely harmonized. His reforms included a requirement for the nobility to shave their beards (perhaps no small achievement in itself since only a generation before the Patriarch had declared that shaving was a mortal sin), western dress for men and women, and encouraging if not indeed demanding that women be admitted into the general company of men, especially into the salons as was common in Europe. Coinage was improved, standardized, and increased. The Russian Orthodox Church was completely neutralized as an independent force in Russian affairs. Peter took control of its finances, and when the Patriarch died he left the post vacant for years. He eventually abolished it, replacing it with a government appointed Holy Synod and led by a lay official called the Ober-Procurator, a situation which continued till 1917. He reorganized the army along European lines making it permanent and professional (not like the *Streltsy* - or the Turkish Janissaries - who became deeply involved in commerce), he acquired improved firearms and artillery, and made advancement through the ranks, even to the highest levels, on the basis of merit alone.

Education, oriented towards largely practical matters, was encouraged and supported. Merchants were sent abroad to study craftsmanship and trades. The gentry were required to complete five years of schooling and denied a certificate of marriage if they couldn't show such an attainment. Professional military schools and the Academy of Sciences were established, and some valiant attempts made to reduce the ghastly ignorance of the common clergy.

Perhaps the most telling reform was the adoption of the Julian calendar in 1700. Before this the Russian new year began on September 1, and years were numbered from the Creation, believed to be in the autumn of 5508 B.C.. Accordingly, the year 1700 would have been 7208. I say "telling" because it not only exemplifies how appallingly backward Russia was, but how even the reform was wanting. It was as though reform was enacted knowing that reform was necessary or desirable but not why it was necessary or with any understanding of the underlying principles. The Julian calendar, which was adopted by Julius Caesar in 46 B.C., set the solar year at precisely 365 1/4 days. Nevertheless, this was not quite correct so that by the time of Pope Gregory XIII the calendar led the sun by 10 days. A revised calendar (the Gregorian) was adopted in 1582 to 1584 by most of the Catholic countries of Europe and gradually by the Protestant countries beginning with the Dutch in 1699. Thus a modernizing movement to discard the flawed Julian calendar was well under way in Europe just as Peter's Russia was adopting it. (However, some believe he rejected the Gregorian calendar for religious/political reasons fearing his Orthodox countrymen would rebel at dating in accordance with Catholic practice.) The Julian calendar remained with Russia until 1918, hence we often see previous Russian dates stated "old style" (O.S.) and the Russian Revolution of November 7, 1917 called the "October" Revolution for its "old style' date of October 25.

The ambition and far reaching consequences of Peter's reforms were peculiar in other ways as well. They affected only the nobility, clergy, and merchants completely missing the vast majority of the people. And very often, even with those affected, the reforms were strangely superficial; some were grudgingly accepted, while others were accepted as beneficial but viewed as essentially foreign, much like a lifesaving organ transplant that nevertheless requires heroic drug therapy to prevent tissue rejection. In this case the will of the Tsar was the equivalent drug therapy, and when it stopped, rejection followed.

There is no better example of this than the fate of the navy into which Peter put so much energy and hope. As we shall see shortly, the Russian navy was used well for the few years that it existed, but with Peter's death it was neglected and allowed to rot at dockside, a process accelerated by the often poor construction of its ships.

Peter's efforts to improve education in Russia also largely failed. The Academy of Sciences he created had to be staffed entirely with foreigners because there were no qualified Russians. In the beginning even the students had to be brought in from abroad, and for the same reasons, there were no qualified Russians. Indeed, things were so bleak at the Academy that at times lectures were given with only other faculty in attendance and no students whatsoever. As for those sent abroad to study, the main result was confusion and continued clumsy ignorance. The arbitrarily selected students were placed in training irrespective of aptitude or inclination. Thereafter they were frequently moved from one school to another in mid-training thus depriving them of any coherent education whatsoever. Sometimes, with equal arbitrariness, they were simply drafted into the army, again in mid-course. Others sent abroad were not given financial support or clear instructions as to what they were to do; the result for students who could afford it was a casual sojourn indulging in the good life of idleness and a return to Russia as stupid as they left. Schools at lower levels were founded but had only marginal financial support. The buildings provided were often dilapidated and poorly maintained. Teachers were hard to find and keep, and the students often fled. The village clergy were perhaps more impervious to education than any other element of society. Suspicious and indifferent to learning anything new made them a hopeless crowd. They simply passively awaited the Tsar's inevitable death and reverted to their old ways.

The rarity of educated Russians did have one important positive consequence, it forced the nobility to use foreign tutors for their children. Eventually this resulted in a nobility with foreign tastes, often French or German, as well as some knowledge and sophistication previously absent from Russian culture. This provided a growing link to the West, but it also further distanced the nobility from the peasantry.

Church reform was successful as far as it went, but it essentially missed the point. Though Peter clearly intended to make his power absolute by absorbing and neutralizing the Church he showed little sense of the power of passive resistance in the peasant's attachment to the faith independent of the Church. (A possible exception might be the avoidance of the Gregorian calendar reform.)

While Peter was practicing war in the Crimea, traveling in Europe, and beginning his Russian reforms, Sweden was positioning herself to have a substantial impact on Russian history by wanting to keep the Baltic as her private sea. Sweden's control of much of the Baltic coast and all but one of the river mouths emptying into it, directly blocked Peter's ambitions for access to the sea and a short trade route to Europe. Accordingly, in 1700 the stage was set for Peter to get into the exhausting 21 year Great Northern War. Initially he and his allies expected to rather easily defeat the 18-year old Charles XII, but they quickly discovered that Sweden's young monarch was a military genius. In Estonia, at the Battle of Narva, Charles' audacity, imagination, energy, and personal leadership combined with the complete incompetence of the much larger Russian army resulted in the defeat of the Russian force so thorough that Russia was ridiculed throughout the courts of Europe. Thereafter, Charles busied himself with an obsessive pursuit of foes in Poland and elsewhere while Peter enacted his military reforms and rearmed his troops with imported modern arms. By 1702 Russian counterattacks yielded some minor but important victories around the mouth of the Neva River. Immediately Peter determined to build a military fortress and city there to protect his long desired access to the Baltic. When Charles heard of the construction he was so sure of himself he said that whatever the Tsar built he (Charles) could take it whenever he liked. In fact, neither Charles XII nor anyone else has ever taken St. Petersburg from Russia.

St. Petersburg was founded in 1703 with the construction of the Peter and Paul Fortress. Very quickly a wooden city followed. In a typically Russian fashion Peter used brute force and labor intensive methods in place of skill and the most basic tools to fill the marshy and swampy site. Laborers literally dug with their hands and carried dirt in their shirttails. The inhuman situation cost the lives of considerably more than 20,000 peasants and prisoners (and some say as many as 100,000). But Peter got his city. He intended it by turns to be a military outpost, then a commercial center, then his capital. In 1708 the nobility were required to move there, and in 1712 he made it his capital. However, it was still a wooden city and so poorly made that even the best of residences suffered from leaking roofs and cracked walls and floors. Accordingly, in 1714 Peter decreed that construction of stone buildings was prohibited everywhere in Russia - except in St. Petersburg. This forced all the country's stonemasons to come there and rebuild the city in stone. By 1717 the city existed more or less complete, and thereafter it grew through natural evolution, though stone buildings remained uncommon even up to Catherine the Great's time.

Meanwhile, Charles had decided it was time to finish off Russia. He invaded heading straight for Moscow in 1708. In a sequence of events that was to become familiar twice more in the next 135 years the invading army was swallowed up by Russia. First Charles had supply problems exacerbated by Russia's scorched earth policy, then he disastrously changed direction and headed south for the Ukraine. Eventually he was thoroughly defeated at the Battle of Poltava in 1709 (Map 6). Though this defeat didn't end the war (Charles escaped), it did establish Russia as a major European power. The war dragged on with the most notable event being Peter's brilliant use of his new Baltic fleet that included a victory over the Swedes at the Battle of Hango in 1714. When peace was finally achieved in 1721 a jubilant and greatly relieved Russian people bestowed the title of "Emperor" and the sobriquet "The Great" upon Peter. Sweden settled by giving up to

Russia the east end and southern shore of the Gulf of Finland, the Estonia and Latvia of today.

During the Great Northern War there was also something of a distraction to the south with the Ottomans. After Charles' escape from Poltava he convinced the Sultan to declare war on Russia and retake what had been lost at Azov in 1696. A rather overconfident Peter responded, but he and his army were surrounded in 1711 at the Pruth River in Moldavia, and he had to negotiate their freedom. It cost Peter all of his gains from his first victories against the Ottomans 15 years before.

When Peter the Great died in 1725 he left behind a state completely free of debt. This is all the more remarkable considering the costly enterprises of his reign; he built a modern army and navy, he fought the hugely costly Great Northern War, and he built St. Petersburg. A comparable achievement for the United States would have been concluding World War II debt free! The catch is that this wonderful accomplishment was largely paid for by the indescribable suffering and labor of the peasantry.

Peter's success was incomplete. He left behind a nobility that was only just beginning to transform from its crude, vulgar, medieval character to a cultured, urbane sophistication comparable to that of the salons of Western Europe. With trade and cultural contacts growing by the decade Russia was no longer a land of complete mystery. Moreover, she had now become an important player in European political affairs. Nevertheless, just as one should not confuse Russian religious beliefs and practices with the Russian Orthodox Church, one should be careful not to confuse the culture of the Russian elite (which typically amounted to less than one percent of the population at the time) with the culture of the peasantry. The peasantry remained as backward as ever. Here we begin to discern, perhaps for the first time, the emergence of two Russias, one modern, one backward; one open and understandable, one not only closed and enigmatic to the outsider, but enigmatic to the Russian elite as well.

The next half century of Russian history is largely a tale of further westernization, of consolidation of power for the crown, and of the growing independence of the nobility. Russia was frequently involved in European alliances that allowed her to fight Sweden and Turkey again, but the results were insignificant. She

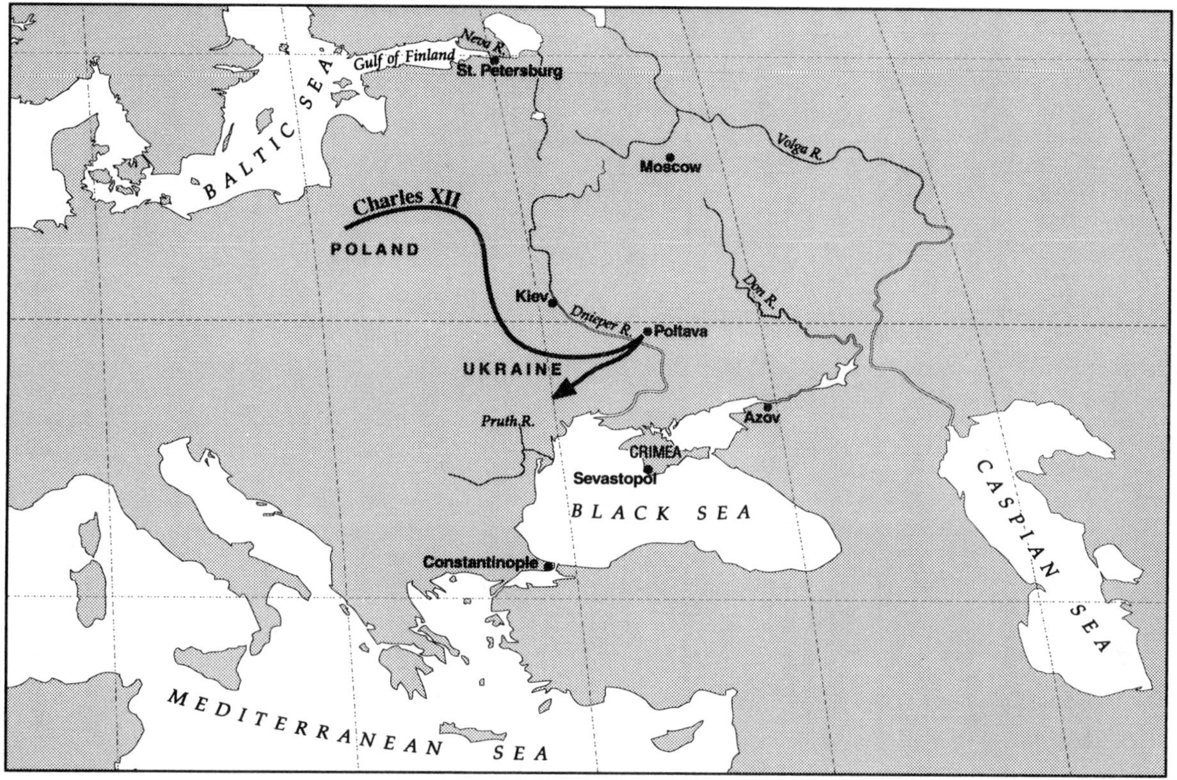

Map 6
Peter the Great's Time

had her first Empress (Peter's second wife, Catherine I), and once the precedent was set, three more, each of whom attained their position with more or less help from the Imperial Guards. The last of the three was Catherine the Great.

Catherine II came to power in 1762 by persuading the Imperial Guards and the discontented nobility that she would be better for Russia than her husband, the infantile Tsar Peter III. A measure of her persuasiveness and unusual capability was that she was German by birth and was deposing the grandson of Peter the Great. Her success was due in no small part to her 20 years of careful preparation. When she came to Russia as a young girl to be the wife of the future Peter III she took the trouble to convert to the Orthodox religion and to learn the language and customs. Furthermore, her ascension to the throne was made easier to accept because the previous 37 years had seen three other Empresses enthroned. Poor Peter was, alas, murdered shortly thereafter, his reign lasting barely six months.

Peter III's brief reign was remarkable for two reasons; his utterly childish behavior in both personal and state matters, which led directly to his downfall, and his abolition of compulsory state service for the nobility. Thereafter they were no longer in literal bondage to the state as the serfs had been and would continue to be for another century. This gave the nobility the freedom and leisure to attend to their estates leading eventually to a class independent of the interests of the crown. Moreover, the leisure and further exposure to Western ideas and education began a process that would eventually spawn another class with independent interests, the "intelligentsia." The consequences of this class fracturing would lead to a historic transition for Russia beginning near the end of the next century.

If Peter the Great brought Russia to the door of Europe, Catherine the Great dropped Russia soundly in its living room. The Russian nobleman at home and traveling abroad was no longer the crude, narrow, arrogant boor of his recent forbearers; he was becoming more educated and sophisticated, not only in the arts but in politics and international affairs as well.

Education was key albeit painfully halting. Early in her reign Catherine created two women's schools admitting both nobility and townsfolk, although in 30 years there were less than 1,000 graduates. She issued several orders and decrees to establish schools throughout her reign but nothing ever came of them. Other plans for universal education were drawn up, some with modest success, yet by the time of Catherine's death only about half of Russia's towns had a public school, and once again with all the good intentions in the world only a very few peasants were enrolled. The total number of students attending was less than one percent. Nevertheless. even with this mixed beginning, and the virtual exclusion of the peasantry, literacy and the general cultural level otherwise were clearly improving as reflected by the substantial increase in the number of periodicals and books published.

Catherine's most visible contributions to the general culture of Russia can be seen today in St. Petersburg, where the city's architecture was transformed from wood to stone, particularly in the palaces of the nobility and the buildings of state. She also created the Hermitage museum which eventually acquired a collection of masterpieces that now makes it possibly the greatest art museum in the world. Elsewhere in St. Petersburg, down river half a mile from the Hermitage, Falconet's monumental equestrian statue of Peter the Great stands as a tour de force in the art of bronze casting as well as human labor.

In addition to being a superb and innovative work of art, the statue and its stone foundation were conceived from the beginning in truly monumental terms. The granite monolith on which the 17 foot statue stands is 16 feet high and weighed 3 million pounds when moved from its natural site 12 miles from St. Petersburg and before being cut to its present shape. The task to move it took 400 men two years. Because of the statue's shape and size casting the bronze required a unique and difficult technique, and only after several failures was success achieved in 1782. That the statue was more than simply a tribute to Peter the Great is clear from the inscription Catherine put on it, "Peter the Great" and just below in letters of equal size, "Catherine II." The inscription itself is in Cyrillic Russian on one side of the statue and Roman Latin on the other. Catherine after 20 years as Empress of all the Russias was declaring that she was the equal of her hero, and she was declaring it to the world. The artistic triumph hailed the political triumph.

Earlier, Catherine had been making good faith efforts to

reform the chaos of Russian laws. She established a Grand Commission calling for representatives from all classes but the serfs. Unfortunately many from the lower levels declined to serve out of suspicion, despair, or laziness. Other elements of Russian society responded more eagerly, and in a burst of gratefulness, began their deliberations in 1767 by proposing a new title, "the Great," for Catherine. However, after two years of meaningless discussion the Commission was dissolved without producing anything. It was a failure; the Russian people were not comfortable with the idea of thought independent of the sovereign.

So the political triumph was not without its difficulties. Catherine's humanity and liberal views did not include freeing the serfs, indeed, her many gifts of land and serfs, as well as her further chartering the oppressive nobility, simply increased the serf's numbers and suffering.

A symptom of the new level of repression of the serfs under the freed nobility, as well as the inhuman treatment of those now forced to work in the emerging industrial sector, was the 10 month Pugachev rebellion. The Don Cossack Emelyan Pugachev led a peasant revolt in 1773-74 that enveloped a huge area between the Ural mountains and the Volga river and for a time put Moscow in jeopardy as well. Other rebellions had occurred before, but this frightened Catherine by its extent and Pugachev's claim to be Peter III. Many simple, willful peasants accepted the pretender's claim, thus the revolt sought legitimacy and aimed directly at the throne. Though Pugachev's demise was reassuringly quick, the rebellion helped transform the liberal and enlightened views of the young Catherine. The popular, regicidal revolution that appeared a few years later in France finally hardened her views to conservative orthodoxy, and the lot of the serfs became as oppressive as at any time in Russian history.

In international affairs Catherine's Russia was becoming a player of some consequence. Russia's historic ambitions and growing involvement with Europe led to two wars with Turkey. Her success was an indication of what was happening on both sides of their power balance; Ottoman power was declining from the heights occupied as they stood before the gates of Vienna in 1683 to the level a century in the future when the Empire would be tottering on the brink of oblivion. On the other hand Russian power was definitely in the ascendent. For the first time Russia consistently beat the Turks on land and sea so that by 1792 she possessed half the north coast of the Black Sea including the Crimea. Peter the Great's dream had been realized, and Russia had access to the Black Sea for the first time since the 11th century.

Closely related to her successes against Turkey was Russia's absorption of parts of Poland. As a weakened Poland tried to secure its sovereignty, Prussia, Austria, and Russia in three successive partitions between 1772 and 1795 completely absorbed it. Russia finally recovered the remaining parts of western Ukraine and Belorussia lost to Lithuania several centuries earlier during the Tatar Yoke. And in what must have been a particularly satisfying added bonus she got that part of Poland that was formerly her ancient enemy, Lithuania.

A relatively minor episode in Russian history is worth telling here because it illustrates an important Russian characteristic that one sees again and again. The recovery of the Ukraine and the Crimea did not eliminate the Turkish threat in the south. At the same time the effects of the Mongol conquest half a millennium earlier remained evident in the low population of this vast region. Catherine and her most competent and influential minister Gregory Potemkin recognized that such an empty space made a poor barrier against the Turks. It would be a good idea if the region could be developed, people brought in, villages and roads built, crops planted, and general commerce encouraged. Not only would success in such a plan provide the desired barrier to the Turks but it would be a magnificent demonstration to the rest of the world of Russian strength, energy, and progress. Catherine gave the job of developing the southern Ukraine and Crimea to Potemkin who later reported amazing progress.

In response to the glowing descriptions of a bustling steppe land Catherine undertook her Great Tour in 1787. She would see the success for herself, and she would bring with her the St. Petersburg diplomatic corps as well as the King of Poland and the Emperor of Austria. All would see Russia triumphant. There was a slight problem however, Potemkin had promised more than he could deliver. While there was no question that much had been achieved, there was also no question that the Ukraine had a long way to go to reach Potemkin's glowing descriptions.

Potemkin's solution was to substitute the impression of great success for its substance. A major part of Catherine's route was a well defined float on sumptuous barges down the Dnieper River from Kiev. Periodically along the way she and her guests would see crowds of saluting peasants and fresh villages adorning the shore. It was a stirring and reassuring sight, and it was at least in part quite false, for Potemkin often had the same crowd of peasants transported down river over night while the royal party was at anchor. The next day's progress would reveal more peasants cheering from the banks. Who would suspect that they were the same peasants as seen the day before! As for the village houses themselves, many were simply facades. One keen-eyed member of the party wondered if the village houses had roofs or inhabitants. Thus was born the notion of the "Potemkin Village," a facade or deception. When one wishes to cover a weakness, one creates a Potemkin Village. (Some historians, most notably Russian, claim the story was a malicious lie fabricated by enemies of Potemkin and the Russian people.)

When Catherine the Great died in 1796 she left a Russia that was profoundly transformed at the upper levels of society. Her work nicely complimented the beginnings made by Peter the Great at the opening of the century. The operation of government was improved and made more efficient. The military was brought up to date. She had subdued all her closest enemies and no nation could seriously threaten her. Geographically Russia was now firmly upon the Black and the Baltic seas. Education was improved, and integration in Western political and cultural life was nearly complete. And Catherine made the nobility more independent than ever, but in so doing considerably increased the burden on the peasantry.

While an overview of Russian history shines with the exploits and achievements of the "Great"s the matters of everyday life for the overwhelming bulk of the people were anything but shining. In Catherine's time as in that of Peter the Great and Ivan the Great before her, the peasant labored, paid taxes, and died exhausted for the glory of the state, but he rarely did so willingly. The peasant often fled the tyrant and tax collector by heading south or east. In the east he was usually too difficult to find or control. However, in the south he and the land were too valuable to ignore. Imperial wars and projects required his money, his muscle, and his life. Inevitably, the suffering masses demanded justice, and rebellion would break out. Almost every Tsar faced these peasant revolts. In the 19th century they would become nearly unmanageable and in the 20th they would destroy the monarchy.

Grand historic figures are sometimes better understood as more the beneficiaries than the makers of history. Catherine the Great was both; certainly her achievements against Turkey would not have been possible with Turkey as it had existed a century or more before. She benefitted immensely from the Ottoman decline. Similarly she was simultaneously pushing her country to the West and being drawn there by events and processes largely beyond her control. She feared, or said she feared, the influence of the revolutionary movements boiling in France and spreading to Poland. This was her rational as she took more measures to control internal affairs. However, her 1793 invasion of Poland further polarized the various national interests in Europe setting the stage for 20 years of war that would find Napoleon occupying Moscow and then Alexander I parading in triumph in Paris.

Chapter 9

Tsarist Russia: the 19th Century

With Catherine's death at the threshold of the 19th century the imperial throne of Russia was briefly occupied by her ineffectual and temperamental son, Paul I. His main contribution to Russian history was to help begin the reaction to European influences that was dominating the elite culture and that would become more coherent and explicit later in the century under the Slavophil movement. He was assassinated in 1801 when his behavior became threatening and insufferable to those around him. Of the six Tsars including Paul who would rule Russia till the end of the Romanov dynasty in 1917 three would be assassinated and one die in his prime "unexpectedly;" the other two would die in their fifties. The Russian imperial throne, for all of its potential power, was not a very safe seat.

By the time Paul's 23-year-old son became Tsar Alexander I Russia's entanglement in the political affairs of Europe was complete. In the beginning England, Austria, and Prussia allied with Russia against France on the one hand and Turkey on the other. But this didn't last long, for alliances were continually shifting with the natural intent of allowing no one power to become dominant. In the middle of all this was Napoleon - and as a kind of side dish, Russia fought Sweden and acquired Finland.

Russia's military prowess was not as impressive on the battlefield as it was believed to be in the salons and ministries of Europe. At the battles of Austerlitz in 1805 and Friedland in 1807, among others, Russian armies were just more cannon fodder for Napoleon's guns. The Treaty of Tilsit in 1807 between Alexander and Napoleon gave each side some respite but was doomed to eventual failure as their respective interests were fundamentally different. This set the stage for one of the two most important Russian wars of the 19th century.

In June 1812 Napoleon invaded Russia. In spite of supply problems and bad summer weather his 500-mile penetration went relatively easily, the Russian resistance being insignificant until the Battle of Borodino 70 miles west of Moscow on September 17, 1812. Of the 250,000 men engaged that day fully 100,000 were casualties; 40 percent! The forces and casualties were about equal with the French numbers slightly greater in each category. Though the battle was essentially a draw, the Russians withdrew and Napoleon entered Moscow a week later. Lack of supplies forced the French to retreat in the autumn of 1812. The Russians still declined to face Napoleon in an open fight, preferring to let the cold, distance, and starvation do the work. By the time the French got to the Berezina River north of Kiev they were weak enough to be directly attacked in force and largely destroyed. Less than five percent of the 600,000 invaders eventually escaped. The whole episode was eerily reminiscent of Charles XII's adventure a century before. (For those interested in something beyond the relatively dry historical narrative, a taste of the times and people is put brilliantly before us in Leo Tolstoy's incomparable novel *War and Peace*.)

Russian armies continued in pursuit of Napoleon picking up allies as they went. In October 1813 Russia, Austria, and Prussia decisively defeated Napoleon at the Battle of Leipzig and triumphantly entered Paris the following March. The Tsar, and with him all of Russia, stood at the center of Western Civilization hailed by all as their savior. It was a heady moment perhaps unmatched in Russian history before or since (with the possible exception of Nazi Germany's fall in 1945).

The War of 1812 was important because it left Russia the dominant power in Europe. It was also important because it brought a very large number of Russians in the officer corps into extensive contact with modern European political ideas that were to have consequences in the Decembrist conspiracy of 1825. Also, the triumph over Napoleon inspired Russian pride and self confidence as well as renewed faith in the Tsar as the special representative of the "God of Russia" and leader of "Holy Russia." These messianic notions, always latent in the Russian people, were reawakened by the invasion threat and used in part to counter Napoleon's early promises to emancipate the serfs. Messianism is one of the deep currents in Russia that seldom surfaces but profoundly influences her history. We shall see it more than once again before we reach the end of our story.

Alexander I died somewhat mysteriously in 1825. Some thought it might have been suicide, but many more believed the rumor that he secretly abdicated to become a monk. He was a very religious person, and indeed, the Russian Senate, State Council, and Holy Synod together conferred on him the title of "The Blessed" in appreciation for his victory over the French as well as for his piety. Though there is some evidence to support the abdication story history generally accepts that he died on December 1, (November 19 O.S.) 1825.

There was some confusion as to the proper succession at the moment of Alexander's death. His eldest brother, Constantine, had secretly renounced the throne some time before. So when the younger brother Nicholas was proclaimed Tsar it wasn't clear what was going to happen. At this moment a group of young nobles who had been meeting since 1816 decided to act in what has ever since been called the Decembrist Revolt. They called on Constantine to be Tsar (hopefully eventually to become a constitutionally sanctioned monarch) and otherwise resisted the elevation of Nicholas. Under the influence of European ideas and a general sense of moral outrage they opposed the historical autocracy of the Tsar and wanted the abolition of serfdom and concomitant redistribution of the land. A poorly led revolt ensued that was easily put down by Nicholas. There followed several thousand arrests, and five of the leaders were hanged with many hundred others sent into exile. The net result of this otherwise insignificant moment was a very substantial increase in state intervention in private affairs with much spying, censorship, and interference in education. Though unrecognized, it was also a significant hint of things to come.

The reign of Nicholas I was notably reactionary. As military and autocratic as any Tsar ever, he and his advisors wished for nothing more than to maintain the existing order of monarchy at home and throughout Europe and the Near East. A ruler was a ruler, and revolutionaries should not be tolerated. Unfortunately for

Nicholas revolutions were occurring all over Europe. Most of the time it was straight forward to be against them, but some times the choice was more difficult as when his Orthodox brethren, the Greeks, wanted to rebel and throw out their Turkish rulers. So, reactionary or not, other circumstances were important as well.

Russia was interested in expanding into the Balkans and, with the ambivalence noted, in helping the Greeks fight for their independence. Accordingly, with a little help from the West the Russians helped the Greeks get their independence. Russia focused on the Balkans and prevailed after a minor war with Turkey acquiring some territory indirectly or outright including Moldavia. She also got the unhindered commercial use of the Turkish Straights out of the Black Sea. Territorial ambition and the natural desire to defend the Orthodox Church against the infidel Turk were satisfied for the moment.

But by now Turkey was no longer a threat to anyone. It was cuffed about, but it wasn't allowed to die as it was generally propped up to be a kind of neutral buffer state astride the gateway to the Black Sea. Later developments in the early 1830s found Russia taking the role of 'protector' of Turkey and gaining out of that the exclusive right of passage through the Straights. France, Britain, Prussia, and Austria would have none of it, and in 1841 got Russia to accept a joint guarantee of free passage. However, things continued to get worse as the Western powers feared Russia's growing influence over Turkey. Diplomatic efforts at reconciliation failed, and war became inevitable. (It was at this time that Nicholas I remarked to the British ambassador regarding Turkey that they had a sick man on their hands, whence the paraphrase, "Sick man of Europe.") In 1854 the confrontation and shocking disaster of the Crimean War hit Russia with the Allied landing near Sevastopol.

Few wars have been fought with greater mutual incompetence than the Crimean War. Allied planning was virtually nonexistent, there were no maps nor even the vaguest idea of the terrain. In the beginning all that was available were a few pictures and drawings, rather like tourist picture post cards. One proposal was to isolate the Crimean peninsula at its narrow base by bringing the isthmus under naval fire; a nice plan until someone discovered that the sea in the area was only a couple of feet deep. A further idea of the

British inept leadership can be sensed by Tennyson's *The Charge of the Light Brigade* romanticization of the stupid cavalry charge that killed or wounded 40 percent of the 700-man unit in 20 minutes. Elsewhere throughout the campaign, disease was rampant and hospital care atrocious with 85 percent of war deaths due to disease and poor medical care. It might have been worse but for the intervention of Florence Nightingale and her nurse corps. It was like the lowest kind of slapstick comedy, except for the death, injury, or sickness of more than half a million men.

The seriousness of Russia's problems can be gauged by the fact that she was even more incompetent than the British, French, and Turkish Allies. Unlike the Allies the Russians had maps, but they were so poor that on one occasion a Russian regiment marched for an entire day only to find itself in the evening right back where it started. The Russian transportation infrastructure was so bad that two-thirds of the recruits sent to the Crimea died en route. The only railroad of consequence in the country was between Moscow and St. Petersburg. (Earlier in his reign Nicholas, responding to a proposal to build railroads, remarked that he could see no use for them. One wonders if he got the picture by 1854.) The roads were virtually impassable throughout the country with command and control from remote Moscow or St. Petersburg essentially nonexistent. The Allies, however, had a direct telegraph connection from the Crimea to London and Paris, and were better able to reinforce and supply their troops over a 4,000-mile sea lane than Russia was over 800 miles of their roads.

Russia's commanding general believed in the maxim of his mentor from the previous century, Suvorov, that the bayonet was better than the bullet. Accordingly, the Russian army was armed with short range, inaccurate smoothbore muskets. The Allies had the rifled musket firing a conoidal bullet of exceptional range and accuracy. In fact it could outrange the smoothbore artillery of the day and still be accurate. In the right hands, as it would be shortly in the American Civil War, it made artillery almost obsolete because their crews were at great risk from sniper fire.

Within a year of Nicholas' death from pneumonia early in 1855 his son Alexander II was ready to talk peace. The Treaty of Paris was concluded in March 1856 ending a war that demonstrated to Russia once again its tendency to lag the West.

The defeat was an utter humiliation with the West compelling terms that included a complete dismantling of Russia's military presence in the Black Sea. Russia had been enjoying the reputation of a major European power ever since Alexander I had entered Paris in triumph 42 years before. Now they found that they were unable to successfully confront the West even on their own land. What was wrong, what was the cause of this disaster?

It was no mystery to Alexander or his people. Russia lost, because her transportation system and especially her roads were an abomination even by the standards of the time. She lost, because her weapons were inferior to the West's. Most of all, she lost, because her troops were physically weak and completely unmotivated. Everyone knew that the reason for this sorry state of the peasant army was directly due to the enervating institution of serfdom. The shock of the Crimean defeat and the clear vision of Alexander II provided the final push to the emancipation of the serfs in 1861.

The idea of emancipation did not come out of the blue however. For half a century it was becoming increasingly obvious that the institution was not only bad but counter productive. Revolutionary movements such as the Decembrists as well as a significant fraction of the nobility were against it on moral grounds. Scholarly and government studies all showed that serf labor was uneconomical. The frequency of peasant revolts was growing exponentially from an average of 15 a year in the 1820s to an average of 78 a year just before emancipation. All in all, by the 1850s it was generally agreed that humanitarian, economic, and political considerations overwhelmingly dictated that serfdom must go.

Naturally many serf owners had reservations on the matter and dragged their feet for several years after the subject became a legitimate topic of public discussion. However, Alexander persisted and with admirable skill cajoled, recommended, and acted until the momentous event became a reality in 1861. Unfortunately as this historic moment unfolded, joy in the principal was muted by disappointment in the reality. The reality was that emancipation was a complicated affair; should all serfs be freed at once? Should they be given land or not, and if so, how much? How were the serf owners to be compensated, if at all? How should all of these matters be treated in various geographic and social areas where land and serf obligations were profoundly different? Some sense of the complexity of the issues can be seen from the fact that the emancipation document itself was 400-pages long.

The freed serfs were not at all happy, because emancipation didn't automatically bring ownership of the land they'd been working. "We belong to the master, but the land belongs to us" was a folk truth for the peasants. And when they did get land they often didn't get as much as they thought they deserved, and they had to pay for it. The terms were generous, but often remained beyond the peasant's ability to satisfy. Moreover, in many instances the land was controlled after emancipation by peasant communes called *mirs*. The authority of these communes, in some ways like the collective farms under Soviet socialism several decades later, wasn't much better for the individual than that of their old landlords. And emancipation didn't free all of the serfs at once; some had already been freed many years before, especially in the west. Others, particularly in the Caucasus region weren't freed for several years more, and others yet weren't serfs but state peasants who were separately provided for in 1863. Emancipation only made the former serf partially free, for there were still tax and other economic restraints as well as the newly increased power of the village communes.

It is a truism that the most unstable time for a society is not the time of oppression, but the time just after oppression is relieved. The most successful Tsar in liberating his people fell victim of this principle. The serfs were emancipated, legal reform more or less made everyone equal in the courts, the Decembrists had been pardoned and allowed to come back from Siberia, and restrictions on personal freedom imposed by Nicholas I had been lifted. Nevertheless, by the 1870s Russia was aboil with new and revolutionary ideas, not the least of which were the questions of authority and Russia's place in history. As these things usually go, Alexander II became the target for seemingly every off-the-wall revolutionary group. And target indeed; in 15 years beginning in 1866 there were at least five attempts to assassinate him. One blew up his dining room, the Tsar saved by an unexpected moment's delay in coming to table. In 1881 a group of "nihilists" succeeded when the Tsar left his carriage to assist members of his entourage wounded by a thrown bomb. A second bomb was thrown mortally wounding him. Thus ended any real chance of a peaceful transition to constitutional monarchy which had been whispered about during the preceding decades. Now the inevitable reaction set in that would carry Alexander III and his son, Nicholas II to the end of the Romanov dynasty and the beginning of the Soviet era.

However, the 19th-century evolution of Russia toward the revolutionary events of the 20th century encompassed much more of significance than war and the rise and fall of Tsars. Russia's geographic expansion rounded out the edges of her empire with largely opportunistic expansion into virtually empty or weakly controlled regions. In the 1829 Treaty of Adrianople she annexed from the Turks the remainder of the north coast of the Black Sea east of the Crimea. The Caucasus area between the Black and Caspian seas including what are now the countries of Armenia, Georgia, and Azerbaijan were absorbed in stages between 1801 and 1878. With more ease yet between 1865 and 1895 she absorbed more than 2 million square miles of central Asia east of the Caspian Sea. This area now comprises the countries of Kazakhstan, Uzbekistan, Kyrgyzstan, Turkmenistan, and Tajikistan. These eight acquisitions worried the dickens out of the British, always touchy about the security of her Indian empire to the south, but no serious problems developed, probably because each empire had just about consumed its limit and had little stomach for more.

Russia was also in its final expansion in the Far East. From 1858 to 1860 she annexed a sizable piece of north Chinese territory including 1,000 miles of coast on the Sea of Japan. This completely blocked Chinese access to that sea. In 1875 she got the large island of Sakhalin in trade from Japan for the Kuril island chain. Together these acquisitions from China and Japan comprised over 100,000 square miles. While she was thus engaged on the west side of the Bering Strait Russia acknowledged an overpowering fact of geography on the east side and sold Alaska to the United States for 7 million dollars. Interestingly, a glance at a world map and a moment's thought about geopolitics raises the question of how long China and Japan, the ascendent nations of the Far East today, will wait to reclaim the territories lost to Russia in the 19th century, the facts of geography and geopolitics suggest not long.

It was during the 19th century that Russia's transportation and industrial infrastructure first became episodically competitive with the West. For 1,000 years her rivers were the only reliable avenues of transportation. Indeed, in 1800 six hundred thousand men were occupied in pulling barges up the Volga River. (Many may be familiar with the deeply expressive *Volga Boat Song* traditionally sung as the men labored against the river.) Eventually many canals were dug to establish east-west links between the largely north-south flow of the rivers. Roads, however, were and remain, notoriously bad in Russia. In the spring and autumn they were impassable muddy tracks, while in the summer they were rutted beyond imagination. Only in the winter were they of practical use as Russians switched wagons for sledges and enjoyed the relatively smooth ride seemingly unconcerned by the very real threat of a frozen death if anything went wrong. Paved roads were unknown until the 400-mile Moscow to St. Petersburg road was completed in 1830 after 14 years under construction. Even decades after the Crimean War, which had highlighted this weakness, the inferior road system continued to strangle industrial growth and was responsible for the frequent wide swings in food prices as farm and city were effectively isolated. In 1893 there were less than 8,000 miles of paved roads in all of Russia, the largest nation in the world. This was one third that of England and one twentieth that of France at the time.

The railroads fared little better. Though the first track was laid in 1835, just five years after the historic opening of the first passenger line in Britain, 20 years later there existed only 647 miles with most of that on the Moscow - St. Petersburg line. About that time there were 30,000 miles of track in the U.S. To this day the Russian railroad infrastructure has never approached Western levels, though the epic achievement of the 5,800-mile Trans Siberian Railroad begun in 1891 is an important exception.

Russian industrial growth in the 18th century was attributable to the plentiful supply of serf labor. However, by the mid 19th century serf labor was no longer cost effective and the work was done by the free peasantry. These were attracted to the Moscow and Ural mountain areas where industry was centered. In 1861 less than one percent of the population of 74 million were factory workers. By 1914 the fraction had risen to just under two percent. As the number of factory workers increased, as well as the size of the factories, the rural peasant workers began to collect in the cities and industrial areas. This brought on an isolation from the land that had provided them with a strong spiritual anchor for 1,000 years. Before long a new class had evolved, the factory worker, otherwise known as the proletariat. As we shall see shortly, it was the small minority proletariat that would provide the revolutionary momentum in Russia, not the majority peasant still on the land.

Russia's 19th-century industrial base was largely centered on iron production and textiles. In 1800 she led the world in pig iron production though eventually the dispersal of iron and coal deposits, and the distance to market overwhelmed the poor transportation system. Once again Russia would fall behind so that by 1860 England was 10 times outproducing her. In the 1880s and 1890s her heavy industry briefly returned to modern standards as she imported equipment, engineers, and managers, mostly from the U.S. and Germany. Some of her steel mills were as large and technically advanced as any in the world at that time.

Textile manufacturing grew rapidly through the first half of the 19th century. By the 1850s Russia had the fifth largest number of spindles in the world; generally however, mechanically powered machinery was slow to be adopted.

On the surface of it Russian agriculture appeared to advance significantly during the 19th century, particularly with its growing exports of grain, but this was only due to more land being put under cultivation. Agricultural techniques were unchanged since the middle ages. Little or no fertilizer was used, tools were primitive, the sickle, so famous on the Soviet flag, was used instead of the much more efficient scythe, livestock was decrepit, weak, or sick, and the roads discouraged distribution to market. Yields were the lowest in Europe and unchanged between the 16th and mid 19th centuries.

It was through education that Russia's long, dismal history began to change. What is most amazing is that anything at all came out her educational reforms, for they seemed as halting, flawed, superficial, and incomplete as all the other reforms that were affecting the Russian people. Nevertheless, the efforts were being made, and as we shall see shortly, their products would have a historic impact on the world of ideas and politics.

During Alexander I's reign five new universities were founded. Along with the already existing Moscow University their admittance criteria were changed and made on the basis of academic merit not rank. Unfortunately political control more or less crippled scholarship and freedom of thought, and things got much worse under Nicholas I with the addition of a stifling censorship. The exquisite sensitivities of the censor were not only oppressive, but on occasion comic, as for example when the possible hidden meaning of an ellipse was questioned in a proposed mathematics book. But absolute control in a large and complex state is illusionary. In this case there still occurred now and then opportunities in Russia for a good and truly liberating education. The net result was that more and more men were entering society with the intellectual tools to analyze and criticize the weaknesses of the culture. Some eventually chose to become professional revolutionaries, a profession, one might say, that they originated. Others simply permeated society establishing fertile ground for change, and others yet became the writers that have since left Russia and the world immeasurably indebted to their insight and brilliance.

These high attainments were for the very few, since most Russians living in the 19th century remained illiterate in spite of the government's good intentions. By 1880 there were 22,000 elementary schools in Russia enrolling somewhat more than a million students. (By way of comparison, in 1880 the United States with about half Russia's population had about 10 million in public schools; Great Britain with one-fourth the population had 4 million.) It's not surprising then that by 1900 only one in four could read.

Early in the 19th century a minor class began to appear called the "intelligentsia." They were, as one 20th-century historian described them, verbose, excitable, impractical, high-minded, and ready for quixotic self-sacrifice. At first they mostly consisted of the nobility, but by the 1840s as educational opportunities spread they came mostly from the small middle class. Though they were the educated and thinking class they rarely seemed to be in touch with the pulse of the people, ie., the peasantry. Their intentions were usually honorable as they discussed Russia's situation and especially the evils of autocracy, but as a group they seemed constitutionally unable to act cohesively. The group broadened so that by late in the century it also included a notoriously troublesome group in any era, student activists. Nevertheless, apart from those involved in overt revolutionary activity, the intelligentsia did make explicit and discuss at some length one of the most basic issues that has faced Russia for half a millennium, the fundamental conflict between the Western and Slavic forces for the Russian soul.

The question that the Slavophils and Westernizers began debating in the 1830s and 40s was what should be the relationship of Russia to the West. Accept? Reject? Compromise? The Slavophils believed that Russia had a special mission in history which could only be achieved by protecting her traditional culture from Western influences. This particularly included the sanctity and primacy of the Russian Orthodox Church. They also believed the country needed spiritual reform. They believed in the Tsar and autocracy but thought bureaucratic power should be limited and free speech encouraged. Collective rights were valued over individual rights, indeed, the individualism of the West was seen as culturally corrosive. Typically the Slavophil saw Peter the Great as a disaster for Russia. Slavophils were essentially conservative.

The Westernizers as a group were less well defined, but generally believed that Russia's special mission must be achieved by following the Western historic path. Nevertheless they shared with the Slavophils a deep reverence for the motherland. They thought that Western values and institutions were desirable, that the Orthodox Church should be disestablished, and that autocracy should be done away with. They were often atheists, usually radical, and thought Peter the Great had been very good for Russia. Socialism and Marxism were brought to Russia by the Westernizers, but they were transformed by the imperatives of Russia's special historic development. All this of course would bear fruit after 1917.

The arguments between the Slavophils and Westernizers wasn't limited to publication of stale political tracts. Indeed, the birth and maturation of Russia's brilliant literature in the 19th century was part of this dialog. Lucky for us, for although Russia has had its own unique and peculiar history, the suffering and experiences of its people are to some degree universal experiences.

It is safe to say that Russian literature of the nineteenth century marks one of the crowning achievements in the artistic and intellectual life of human history. Dostoevsky questions in *The Brothers Karamazov* what is worth the suffering of even one innocent child. One is tempted to respond that all of Russia's suffering through a millennium of history is easily a good price for the literary gift she has given the world. Between 1832 when Alexander Pushkin published *Yevgeni Onegin* and 1904 when Anton Chekhov's *The Cherry Orchard* appeared the world's literature was transformed. Wonder at the further appearance of geniuses like Nicholas Gogol, Ivan Goncharov, and Ivan Turgenev, and try to imagine a national literary scene, home at the same time to Fyodor Dostoevsky and Leo Tolstoy! Some have called it the "golden age of Russian literature." One could as easily call it the "golden age of the world's literature." As if this outpouring of literary genius wasn't enough, Russia began producing world class composers as well; Alexander Borodin, Modest Moussorgsky, Peter Tchaikovsky, Nicholas Rimski-Korsakov, Sergei Rachmaninoff, and as the century ended Igor Stravinsky and Sergei Prokofiev. What can one say, except perhaps that here at last the Russian people had found their natural genius and place in the world.

The Russian Orthodox Church and Orthodox religion remained the solace of the peasant, and many merchants and military officers, and some of the nobility in spite of the inroads of atheism. The Church, completely in the hands of the Possessors, continued as a convenient and useful auxiliary for the Tsar. On the other hand, the membership of the Church, largely the peasantry, was mostly Non-Possessors or Old Believers, at least in their beliefs if not their outward practice. The village clergy was heartily disliked, somewhat feared, and viewed as useful only in a formal sense for their services in performing the liturgy. Spiritual ministrations were handled by Elders (*Startsy*) in the monasteries. (There is no better illustration of their role than that of Father Zossima in *The Brothers Karamazov*.)

In the 19th century the Russian Church and religion, as well as the government and the intelligentsia, began to more frequently articulate a special destiny for Russia. It found expression in the "Pan-Slavic" thinkers who claimed it was Russia's responsibility to protect Slavic peoples everywhere. There was a similar umbrella notion for all Eastern Orthodox Christians. These ideas were useful in rationalizing Russia's wars with Turkey. But the vision was greater than this. Russians believed that they had a role in history, that through their suffering and holiness they would lead humanity to a time of love, harmony, and humility before Christ. The writings of Dostoevsky and Tolstoy are particularly concerned with this messianism, and later, ironically, even Lenin and his associates justified their Marxism as this natural fulfillment of Russia's destiny.

However, well before Lenin the government made perhaps the most explicit statement of Russia's messianic role in the doctrine of "Official Nationality." In 1833 Sergei Uvarov the minister of education under Nicholas I proposed the doctrine as a guide for educating Russia's university students. Its three principles, "Orthodoxy, Autocracy, and Nationality," defined Russia and being Russian,me the uniting cry for conservatives for the remainder of the century. Orthodoxy meant that the Russian Orthodox Church was the official Church of the Russian people and the sole and immutable source of spiritual and ethical norms. Autocracy divinely ordained the necessity of the ancient and absolute rule of the Tsar. Nationality recognized the special nature and historic mission of the Russian people. "Orthodoxy, Autocracy, and Nationalism" was a very comforting theme for the Tsar and Russian xenophobes, but by Alexander III's ascension to the throne in 1881 the formula appeared to be obsolete.

While the intelligentsia, the nobility, and the government were indulging themselves in political and philosophical duels the peasantry remained almost a culture by itself. In the 19th century Russia was still a peasant nation since fully 90 percent or more of its people were rural or village inhabitants as physically and emotionally isolated from the powerful and influential of St. Petersburg and Moscow as from Mars.

The 19th-century-Russian peasant has been described by Russian and Western observers as vicious, debauched, cunning, deceitful, closed minded, stupid, slow, lazy, weak willed, self indulgent, stubborn, forbearing, hard working in necessity, long suffering, tender, loving, loyal, and religious. Without a doubt they are all right; this litany, for example, describes Dmitri Karamazov of *The Brothers Karamazov* rather well, but how can any individual or homogeneous group be so contradictory?

The Russian peasant, the epitome of the very essence of Russia, is first before everything else an East Slav. Whatever the forces that defined the East Slavic personality in the centuries before Rurik the character remains clearly visible in the 19th, and to some degree, even in the 20th century. Historians and writers call it the Russian soul.

After recognizing the essential and preexisting East Slavic (or simply Slav or Slavic from now on) personality of the Russian peasant his next most important characteristic is his Orthodox religious beliefs. However, it is clear from the *Primary Chronicle* that the Rus chose Eastern Orthodoxy because it was most harmonious with their already existing personality. They chose Orthodoxy, not the other way around as sometimes happens when missionaries aggressively proselytize a people.

So there we have it in a nutshell; the Russian peasant is an Orthodox Slav. His religion is an amalgam of Christianity and paganism. He believed in a nature well populated with sprites (reminiscent of the Arab Muslim's *jinn*) and the all embracing Mother Earth. He evolved in a region of harsh circumstances and was usually overwhelmed by the peoples around him. In the face of these forces he coped by outward submission and inner resistance. He kept his true feelings from strangers. He respected and feared the gods and felt helpless in their grip. Helplessness was the natural state of things. He also felt even more that suffering was the natural state of things, and that it was the path to salvation provided by God and exemplified by Christ. Accordingly, suffering was good, even desirable, never mind that it might be a consequence of antisocial actions. Thus one observes historically the peasant's almost serene self acceptance of any abominating behavior; not only does one not have to feel guilty about it, but suffering the resultant social sanctions eases the way to salvation.

The sense of humility and acceptance was so strong in the Russian peasant that he looked with suspicion, if not real anger, at those who would presume to change their state and better themselves. Everyone had an assigned job, and it should be accepted; it was bad to want differently. Ambition was suspect. From this it is easy to see how difficult it was for Russia to progress even with the best intentions of the leadership. This also didn't make fertile ground for revolutionaries. For example, student activists in 1873-4 tried to arouse the peasantry but were turned in to the government by these very same peasants. (This is also an example of the gap between the intelligentsia and peasant, for where the intelligentsia considered the Tsar as the enemy the peasant similarly regarded ambition and anything new.) Later Lenin would also dismiss the peasantry in frustration as well. Evidence of this rejection of ambition remains visible today, for example, consider the following joke popular in Russia in the early 1990s: Peasant Ivan is plowing his field and unearths a bottle, which he uncorks. A genie rushes out and, grateful for being freed, tells Ivan that he may have one wish. Ivan's eyes light up. "See that cow over there?", he says, pointing toward his neighbor Vasya's field. "That cow belongs to Vasya, and Vasya has become prosperous ever since he got it. He milks it, makes cheese and butter, and then sells them at the market for a fat profit. It's just not fair." The genie asks Ivan if it is his wish to be given a cow like Vasya's. "No, no, no," says Ivan. "I want you to kill Vasya's cow."

Of course avoidance of change wasn't universal, but it was prevalent. Moreover, as the centuries passed, other historic forces were actively transforming the peasant. Though God and history may have made suffering desirable, it still felt good to be relieved of it. Drink and intoxication worked well, always had and perhaps always would. (Communist General Secretaries and President Yeltsin were still dealing with this in the late 20th century.) And the peasant shift away from the land, Mother Earth, set him adrift and open to new influences. But beneath it all their remained, and remains, the Orthodox Slav.

The new influences on Russia as the 19th century drew to a close were the growing power of Western ideas and materialism. The net result was more revolutionary activity, more government reaction, and more instability. Inevitably then, the reigns of Alexander III (1881 - 1894) and his son Nicholas II (1894 - 1917) were largely holding actions against growing social unrest.

Nicholas II, the last Romanov, was the weaker of the two. He inherited a highly repressive state from his father. In 1880 the "Third Section," a kind of counterrevolutionary police, was abolished for ineffectiveness in the face of terrorism (recall the bomb under the imperial dining room) and was eventually replaced by the Department of Police in 1883 to look after ordinary crime. There was, however, also a separate police called the Corps of Gendarmes. Established in 1881, its job was to protect the state, and it was a law unto itself. It was beyond judicial supervision; it could limit all gatherings of any number of persons, it could close businesses and levy fines up to 400 rubles (a substantial sum in those days), it could arbitrarily detain individuals up to three months, or exile them on suspicion only. Over the next 20 years these measures only further radicalized the students on whom the oppression fell heaviest. One of the government's victims, executed for involvement in a plot to assassinate the Tsar in 1887 was Alexander Ulyanov, the older brother of Vladimir Ulyanov, Lenin.

Imperial Russia had never really recovered from its publicly demonstrated weakness in the Crimean War. Yes, it could still defeat third rate nations such as Turkey in 1877-8, but even here it was forced by the West to reduce the war settlement. In the Far East, where Europe had been having its way for a century or more, Russia's reputation was dealt another blow. In the 1890s Japan began to take exception to Russian expansion and influence over prostrate China. It culminated in 1904 with the Japanese attack on the Russian fleet at Port Arthur. The Russo-Japanese War that followed was the Crimea all over again. Japan thoroughly defeated Russia exhibiting better military competence and organization, and generally possessing superior weapons. It also helped to be fighting in her home waters. After destroying Russia's Pacific fleet and pushing back its armies in Manchuria, Japan then destroyed Russia's Baltic Fleet which had taken the trouble to sail 20,000 miles for the occasion. The Battle of Tsushima Strait in May of 1905 cost Russia 56 of its 59 ships including the capture of the incompetent Russian admiral in command. The peace was concluded at Portsmouth, New Hampshire just in time for Russia to face the 1905 revolution at home.

Chapter 10

The Rise of Soviet Russia

The opening of the 20th century found the weak and incompetent Nicholas II completely incapable of the task before him. Giant Russia was staggering under the forces of change that had been building for 100 years. The rigidity of absolute rule was making itself largely irrelevant as the peasantry was freed of serfdom and began its emigration from village and farm to the urban factory. The Tsar's subjects were developing new values and a certain skepticism of his role. Revolutionaries were preaching the end of Tsardom while others were talking of constitutional monarchy. Meanwhile, abroad, the power of Western thought and industrialization was transforming the world leaving Russia largely behind. The Crimean disaster hinted of the gap. Then in quick succession it was certified by the humiliating defeat in the Russo-Japanese War of 1904-5, the 1905 Revolution, and World War I. Faced with these events Nicholas was as helpless as a child. Inevitably, the spring and autumn revolutions of 1917 followed, and an epoch of Russian history ended. The socialist era was beginning. How much of Russia and its soul survived we shall now see as we sound the force of this tide of history.

One of the direct causes of 1917 can be found in the failure of the Tsar to understand or acknowledge the changes that had been occurring in Russia. Given his character it could scarcely have been otherwise. A charming though private person he was a homebody who enjoyed the intimacy of family life while indulging his ego in a stupid certitude of his divinely appointed mission as absolute monarch of all the Russias. He got some shallow sense of authority and control of events by deeply involving himself in the military minutia of uniform and parade. Public functions off the parade ground, however, generally made him uncomfortable. In reality he never seemed to know quite what to do as affairs of state unfolded before him. This weakness was manifest in both domestic and foreign affairs. He compounded his personal failings with a knack for appointing incompetent ministers. But perhaps the final, most profound, and fatal insult to his people was his excessively Westernized style and taste that severed any emotional link that might have bonded the Russian peasantry to him. The man just wasn't up to the job; but his wife was.

The Tsarina was a very strong willed woman perfectly at home with the compliant and complacent Nicholas. Alexandra had been a princess of the German Grand Duchy of Hesse. Accordingly, her credentials for substantially influencing the rule of Russia through her husband were detested by the peasantry and those higher personages not in power. It was often said that there had been altogether far too much German blood ruling Russia since the days of Peter the Great.

Among her characteristic transgressions was the notorious influence in state affairs she allowed the self proclaimed lay 'holy man,' Rasputin. However, after a decade of suffering Rasputin's crude presence (some believe "Rasputin" was a name attached to him in his youth meaning "debauchee.") and his appointed friends, a small group of conspirators invited him to a dinner party in 1916 and managed to kill him, not without difficulty, by successively poisoning, shooting, stabbing, and then drowning him - all in one long evening.

Together, Nicholas and Alexandra managed to exhaust most of the remaining credibility held by the institution and person of the Tsar in the hearts of the workers and peasantry. This turned out to be as fatal to Tsardom as the poison, bullets, blade, and river were to Rasputin.

The first substantial indication that the monarchy itself was in deep trouble came in early 1905 when a peaceful worker's march proceeded to the Winter Palace in St. Petersburg to petition the Tsar for more political freedom and relief from unbelievably bad working conditions. The crowd was fired upon by the army killing and injuring several hundred. The event, known ever after as "Bloody Sunday," precipitated further worker strikes in the cities and peasant uprisings in the country side. These had been occurring with growing frequency for several years as the life of the worker and peasant had descended to hopeless depths of suffering. Matters grew worse throughout 1905 exacerbated by the war burden and humiliating loss to Japan, the mutiny of the crew of the battleship Potemkin in the Black Sea, and the active involvement of several revolutionary parties that helped change the people's demands for better living conditions to demands for political change. The culmination finally came in October with a massive, 10 day nationwide general strike that forced the Tsar to order the election of a Duma with important legislative powers. The Tsar effectively admitted the end of half a millennium of virtually absolute rule.

Nicholas very quickly found that the elected Duma was too liberal for his taste and dissolved it after two months. A second Duma was elected, but it lasted only three months. Nicholas was surrendering power very reluctantly. Finally, by carefully establishing eligibility criteria, he got a Duma in 1907 he could live with, and it lasted through its full five year term. Imperfect though it was the Duma provided enough of an outlet for popular discontent to ease the crisis of rule for several years. Nevertheless, the strong monarchy remained, as did the fundamental weaknesses in the economy and society, and the discontent of the revolutionary minorities.

The general strike of October 1905 proved to be the occasion for the Social Democratic Party to begin recruiting the workers into highly organized worker's councils, called "soviets." These were natural extensions of the already existing tendency of the worker to organize into unions. It was through the unions that the worker was able to have some influence on his future and reestablish his identity after he'd weakened or severed his traditional ties with his rural peasant past. This was an important transformation for it made him more acutely conscious of and responsive to political values. The appeal was readily satisfied by the cadres of the Social Democratic Party.

The Social Democratic Party was among several that began to form about the turn of the century. After a hesitant start it immediately split in 1903 into factions, one group claiming falsely to represent the majority ("Bolshevik") of its members and the other through ineffective counter propaganda left with the equally false name of "Menshevik" (minority). The Bolsheviks, led by Vladimir Lenin believed that membership in the party should be limited to professional revolutionaries, and that there should be no collaboration or compromise with other political interests. The Mensheviks were more open and willing to collaborate with anyone. Both, however, were Marxist.

During the 1905 Revolution soviets were organized in St. Petersburg by Leon Trotsky and in Moscow by Lenin. They had little affect on events of that year, for the political rigidity of the Bolsheviks discouraged compromise. Indeed, they boycotted the first Duma election and concentrated on further organizing to lay the foundation for seizing power at some future time. However, they did run in the second election and found themselves the third strongest party in that short lived assembly.

It was during the first dozen years of the 20th century that the three principal figures of the Soviet era, Lenin, Trotsky, and Stalin emerged before history. Vladimir Ilyich Ulyanov (Lenin) was born in 1870 into a family of newly designated minor hereditary nobility. All in the family were well educated and intellectual with Vladimir perhaps the most brilliant. He did very well in school with little impact on his grades evident when his older brother was executed in 1887 for revolutionary activity. Nevertheless his radicalism was probably inspired by that event, for the previously apolitical student now went off to the university to study law, quickly became deeply involved in revolutionary organizations, was expelled, and discovered Marx's *Das Kapital*, all in barely more than 12 months. As a non-student he managed to learn enough law to easily pass the law exam in 1891 and practice briefly in the provinces. However, continued contact with Russian revolutionaries throughout this period turned him fully to his historic role. He formally became a Marxist in 1893, was involved in labor organizing, arrested, jailed, exiled, and on release moved to Switzerland where he became very active in writing (taking the pen name of Lenin) and publishing for the cause of the Social Democratic Party. He returned to Russia for the 1905 Revolution but went to Switzerland in 1907 to escape arrest. There he continued to write and organize Russian revolutionaries from a distance until April 1917.

Lev Davidovitch Bronstein (Leon Trotsky) was born in 1879

of a comfortable Russian Jewish family. Like Lenin he was highly intelligent, well educated, and a gifted writer and speaker. At an early age he was moved by the suffering and oppression of the peasant, and in 1897 he became a Marxist committing himself to a life of revolutionary activity as he began to organize Russian workers into a union. As happens, this work got him a trip to Siberia where he passed four years reading Marx and Lenin. He escaped in 1902 taking the name of his jailor (Trotsky) on his forged passport and turned up in London where he met Lenin for the first time. His energy, brilliance, and communication skills quickly raised him to a leadership position within the Social Democratic Party. In the 1903 party split he sided with the Mensheviks, for he believed that Marxism was sufficient in itself to guide and motivate revolution in the worker and that within the Social Democratic Party maximum stability and effectiveness would come from the broadest membership support. This was just the opposite of Lenin's belief in the special and essential role of a group of professional revolutionaries that would lead the less inspired masses. Trotsky expressed his disagreement and misgivings in the remarkably prescient 1904 observation that such an attitude would, "lead to this: the party organization at first substitutes itself for the party as a whole; then the Central Committee substitutes itself for the organization; and finally a single 'dictator' substitutes himself for the Central Committee." Though he later converted to Lenin's view and joined the Bolsheviks, he was always struggling with some sense that democracy within the party was desirable, even as affairs developed according to his prediction. Eventually this fundamental difference would cost him his job, his country, and his life.

The instrument of this fall wasn't Lenin, but Stalin. Like Lenin and Trotsky he had a fine mind, but from there on the resemblance falters. Among other things he was a poor speaker, an indifferent writer, and he was non-Russian. Iosif Vissarionovich Dzhugashvili was born in Georgia in 1879 into rather meager circumstances. His father, a vicious drunk who had been a serf, was killed in a barroom brawl. The son was a rebellious student, though academically good enough to be sent by his widowed mother to the only higher education available in Georgia at the time, the seminary at Tiflis. Contrary to the rules of the seminary

Dzhugashvili read many forbidden books including Marx. Of course this got him in trouble, and after four years the Russian Orthodox Church lost a potential priest, and the Social Democratic Party gained a full time organizer and all around trouble maker. He was so troublesome that the party was going to throw him out, but the police got to him first, and in 1903 he was sent to Siberia. Between 1903 and 1917 the cycle of prison, exile, escape, hiding, revolutionary action, and recapture repeated itself five times. He met Lenin in 1905 and thereafter functioned in compliment with him; Lenin was the thinker and spiritual leader of the Bolsheviks, while Stalin, a pen name Dzhugashvili took about 1912, was the man of action. He was also the party banker, so to speak. Since money through donations was completely inadequate to support their activities, Stalin turned to bank robbery. Lenin also found this active young man useful because as a non-Russian Georgian he was a good link to the other nationalities that might be organized within the Russian Empire.

Marxism and Lenin's transformation of it to accommodate Russia's special circumstances were the rational underpinning of everything that happened in Soviet Russia after 1917. Marxism views history in terms of economics, and man and all of his interactions as essential materialistic. It also views men as either exploited or exploiting. The exploited are called the workers, the proletariat. The exploiting are the owners of the means of production, in modern times the capitalists. History is dynamic; it evolves. According to Marx, this evolution obeys immutable dialectical laws. Marx's dialectical laws say that every historical condition has within it the seeds of its opposite and that eventually the two come in conflict and produce a new condition. (This is succinctly encapsuled in the process, 'thesis, antithesis, synthesis' often associated with Hegel.) Then the new condition itself develops its opposite with the subsequent internal contradiction driving a new synthesis that continues the cycle. Thus the evolution of history, a history essentially of "dialectical materialism" which inevitably produces class conflict between the workers and the owners. Marx thought he saw this theme repeatedly played out through history from the ages of slavery and feudalism to the age of capitalism. He believed that the final synthesis would be the proletarian revolution against the capitalists (the property owners, the middle class were called the bourgeoisie) which would somehow produce a forever stable socialist utopia known as Communism. The workers would own the means of production, so there would be no exploiters or exploited. Everyone would be content in a world driven by the principle of, "From each according to his abilities; to each according to his needs." Even government itself would be superfluous.

Lenin had a serious problem with Marxism as a revolutionary instrument, faced as he was with a Russia that had a minuscule middle class that certainly wasn't up to challenging the monarchy, nor much of a subsequent target itself. Moreover he also had to deal with the vast and largely passive peasantry, and he didn't want to wait for the classical chain of developments; a middle class emerges, overthrows the monarchy, the proletariat emerges, overthrows the middle class, socialism prevails and rapidly transforms into a communist utopia. Accordingly, for Russia he envisioned a temporary alliance of the bourgeois middle class, into which he dumped any peasantry that owned just about anything, and the proletariate worker against the monarchy. The middle class could be disposed of later after power was in Bolshevik hands. The key distinction, however, between Marxism and Marxism/Leninism is based on Lenin's passionate belief that the oppressed masses were too passive to fully revolt; they would just form trade unions and negotiate a settlement with their oppressors. Accordingly, the masses needed to be led to the right path of revolution by a highly disciplined corps of professional revolutionaries, the organizers and leaders of the Bolshevik Party - the Communist Party after 1918. Lenin also feared the more advanced, capitalist countries, so his form of Marxism that was to produce 20th-century Soviet Russia was just as xenophobic as old Russia had ever been. After the revolution Marxist Soviet Russia would then lead the way for socialist revolutions everywhere else in the world. Here we see the coherence of the 1,000 year tide of Russian history manifest even in convulsive revolution; its sense of special mission or messianism, its fear and distrust of the West, and the passiveness of the peasantry.

The stresses of World War I proved to be more than the decaying monarchy could survive. Russian soldiers were sent to the front ill equipped, poorly supplied, and incompetently led to be devoured by the German fighting machine. On occasion they were sent into combat completely unarmed with only a lucky few able to pick up weapons from the battlefield dead. Others were simply armed with long handled axes in lieu of rifles. German heavy guns outnumbered Russian by fifty to one, and in one minute would fire more rounds than the Russians fired in an entire day. After 10 months Russian casualties amounted to 3.8 million killed, wounded, and missing. The Tsar's disappointment in these failures misguidedly persuaded him to assume personal command of the army in August 1915. However, that only made matters worse since the government back home was now in the hands of the Tsarina and her followers.

Political corruption and ineptitude at home exacerbated the problems of worker and peasant discontent. The inadequate infrastructure including poor highways and railroads caused significant food and fuel shortages in the cities, especially in Petrograd (St. Petersburg had been patriotically renamed in 1914) even as it strangled military resupply at the front. Workers through their unions were transforming protests against deplorable working conditions to protests against the very nature of their rule. Russia was now on the edge. A century of clumsy adjustment to join the West as a Power was about to be reckoned.

Economic development of Russia in agriculture and industry in the 19th century was very irregular in time, location, and activity. Occasions of rapid advancement were real enough and impressive when they occurred, but they were characteristically episodic and scattered. Overall, Russia started from a relatively backward position and continued to be relatively backward to the eve of WW I. Her development since Peter's time had always been hugely influenced by the advantages latent in her large population, which could be worked to inhuman levels (and often was), and an immense geographical extent where space and mineral wealth each could contribute to her strength and security (and often did.) All this is as true in the 20th century as it was in the 18th and 19th centuries. But the efficacy of souls and space has its limit, especially when both are largely undeveloped. Unfortunately for Russia, the demands of World War I pushed her beyond the limit; the war became a vast resource sink for a country that had no reserves whatsoever to support it.

The collapse came in two steps. In March (February Old Style) 1917 massive protest demonstrations developed in Petrograd over inflation and the scarcity of food and coal. The army refused to fire on the demonstrators, and indeed, some joined them. The government was helpless and simply collapsed. The people turned to the only authority around, the Duma, which promptly formed a provisional government and called for the Tsar's abdication. By the 16th of March the House of Romanov had fallen and with it the monarchy and Tsardom dating back to the beginning of Ivan the Great's reign in 1462.

Everyone had believed for years that a revolution was coming, but no one was prepared for the ease, speed, and completeness of the fall, not even Lenin in Switzerland and Trotsky in New York. Having been caught off guard, they now quickly returned to Russia to take part in the unfolding events. Trotsky's journey was straightforward and uneventful, but Lenin was stuck in Switzerland, for no nation would allow him to cross their territory to get to Russia. The Germans, however, clearly saw an advantage to having this trouble maker returned to the bosom of their eastern enemy, so they facilitated his journey by allowing him to cross their territory to Russia safely restricted to a rail coach in a "sealed train."

The provisional government, eventually led by Alexander Kerensky, turned out to be unstable and not well liked by the people. Even though it espoused a number of liberal reforms it still seemed somehow out of touch with the people and ineffective in addressing their needs. Moreover, the provisional government favored continuing the war, which by now was becoming unspeakably costly. Meanwhile, another organized political force was making itself felt, the Petrograd Soviet, the workers council organized by the Social Democrats and led by the Bolsheviks and Lenin after mid September. Kerensky, recognizing this threat from the left, tried unsuccessfully to suppress it. The net result was the second and final step in the transformation of Imperial Tsarist Russia into Soviet Russia.

On November 6, 1917[1] the Bolshevik revolution began with the seizure of the telephone system, police and railroad stations, government offices, post offices, and key points in the road system. Similar actions occurred in other towns and cities of Russia, usually without significant resistance, the notable exception being Moscow where some resistance lasted for a week. The provisional government was overthrown in just 36 hours; the "Great October Revolution" was almost bloodless. Life in Petrograd, even during the height of the takeover, went on as if nothing was happening. Thus these momentary actions of the Bolshevik minority passed virtually unnoticed; would that the consequences could have been as unremarkable, for the events of those two days would largely determine the politics of the globe for the remainder of the century. (Kerensky escaped to pass the rest of his life in exile dying in New York City in 1970.)

The initial success of the revolution was due to the good organization and control of the Bolshevik cadre as well as the temporary alliance the Bolsheviks made with the Social Revolutionary Party and with the peasantry. Once in power, however, the Communist Party (Lenin changed the name from "Bolshevik" to "Communist" in March 1918) suppressed its allies, rejected their interests, and proceeded with its own agenda. The eventual achievement would be of epic proportions, but the cost in human lives would surpass anything known in history before or since. Some scholars estimate as many as 54-million Soviet Russian deaths were the direct result of Communist policy between 1917 and 1987.

How did Russia come to 1917? What have we seen in her history and culture that could explain it? And was the October Revolution something way out of line with the past, or were these events in general harmony with the deeper tides of Russian history? One must start with the Russian people and observe that from pre-Kievan times they have been consistently notable for the value they have placed on humility and the salvational efficacy of suffering;

---

1 October 25th according to the Old Style, Julian calendar adopted by Peter the Great. Thus the revolution has come to be known as the "October Revolution." Soviet Russia actually adopted the modern Gregorian calendar on January 31, 1918, and made the adjustment by making the next day February 14.

meekness and selective passivity have been closely associated with the humility. Then, at least from the 15th century, one begins to observe a certain xenophobia, a combination of distrust of foreigners and a peculiar feeling of self superiority. The superiority has been manifest most clearly in Russian messianic traditions from the 'Third Rome' idea and Russian Orthodox religion to Fyodor Dostoyevsky and the 19th-century slavophile's explicit thesis of the special historical role for Russia. These notions then inevitably became attached to Marxism and its messianic undercurrent. The emancipation of the serfs released old forces in the culture to be directed elsewhere. Though the serfs were nominally freed of their masters, they were also to a degree 'freed' of their land. The peasants were quite bitter about the unfairness of having to buy the land they had always occupied and considered theirs. The old serf saying, "We belong to the master, but the land belongs to us" clearly shows the depth of this attachment. Moreover, they were by no means totally free of their masters, either on the land or increasingly in the city as industrialization rapidly developed. So the contract between the worker (and to a degree his country cousin) and his master to silently endure all hardships was being weakened. Add to this the failure of the Tsar, indeed of Tsardom itself, robbing the worker and peasant of hope that the Tsar would or could intercede on their behalf against the exploiters. At the same time Tsardom was also discredited as an agent of God's trials, ie., efficacious suffering, that would further their salvation. Worker and peasant suffering was now from and by men, not from and by God (or His representative). Therefore it was not an instrument of salvation and was essentially pointless as far as they were concerned. A repressive monarchy was historically acceptable to the peasantry only as long as it was seen to be fair or efficacious. All this set the worker, and some of the peasantry, spiritually adrift and made them ready recruits for the first promise of relief and spiritual security to come along. Marxism-Leninism provided that in a congenial form embracing as it did messianic aspects familiar to the Russian people through the centuries of their history. The final catalyst for all of these ingredients was provided centuries earlier when the nobility were freed from court service. This provided a class of citizens with the leisure and drive to get and use Western education. As this trickled down to the lessor classes and as Western political ideas went with it, the organizational apparatus developed to bring these currents together. Thus Lenin, and the Revolution.

One of the first things that Lenin did after coming to power was to arrange an armistice with the Germans in December 1917 and negotiate a peace. He was driven by the specter of a war that was bleeding Russia to death and the belief that the war was a capitalist device to dominate the masses. He wanted peace so desperately that the final Treaty of Brest-Litovsk signed in March, 1918 surrendered very large parts of Russian territory, industry, and resources, including Finland, the Baltic States, and part of Poland and the Ukraine. Fortunately, or shrewdly, as the case may be, the end of World War I eight months later nullified the treaty and very shortly most of the surrendered territory was recovered.

In the meantime, things weren't going well for the Communists; getting power was one thing, keeping it was quite another. The high-handed tactics of the Bolsheviks/Communists in seizing power had alienated large parts of the population including the Menshevik fraction of the Social Democratic Party (which was a majority within the party, name not withstanding), the Social Revolutionary Party (whose representatives outnumbered the Bolsheviks two to one in the short lived popularly elected Constituent Assembly of early 1918), and numerous right wing groups, nationalists, and peasants. The whole mess reminds one of the situation that existed seven decades later as the Soviet Union disintegrated, except that this time civil war broke out.

By the summer of 1918 the various interests excluded from the government by the Communists had coalesced into a resistance force known as the White Russians. At first it looked as though the White Russian army might prevail. Their forces controlled most of Russia except the core region extending for a few hundred miles around Moscow and up to Petrograd as well as east to within 100 miles or so of the Ural Mountains. Less than 10 per cent of Russia was under Lenin's control.

Several White armies operating largely independently began offensives in mid 1919 pressing the defending Red Army from three sides. One army led by a Tsarist general got within sight of Petrograd, but that was it. Elsewhere, invasion of units of 14

Allied countries descended upon the struggling Reds. The most notable were the British and American forces that landed at Archangel and Murmansk, the British and French along the north shore of the Black Sea, and Japanese and Americans in the Far East.

Somewhat amazingly, the Red Army won. After three years of fighting the Allies had largely withdrawn, and the White Russian army was defeated. As usual, such miracles come about through a combination of reasons. The Allied intervention was half-hearted at best, White Army operations were very poorly coordinated, the Red Army was inspired by its revolutionary mission and energetically and superbly lead by Trotsky, and perhaps most importantly, much of the population more and more saw the Whites as counterrevolutionaries working to restore the country and government to its prewar status, and they would have none of that.

The country was exhausted. As 1921 passed Russia had endured four years of war abroad, three years of civil war, a profound political revolution, famine, typhus epidemics, and inflation. One might have easily called it the "New Time of Troubles." Russia, now Soviet Russia, (and in 1922 the Union of Soviet Socialist Republics) was suffering, and if salvation is through suffering, then very nearly every Russian born in the 20th century would have a guaranteed place in Heaven.

When the Constituent Assembly was elected in December of 1917 the Social Revolutionaries won 370 of the 707 seats. But the Bolsheviks were in physical control, so when the Assembly tried to meet in January 1918 they prevented it. From then on it became a time for the Bolsheviks to consolidate their power. But the process was going to be horrendously traumatic.

Draconian measures were instituted not only as things got difficult, but as part of Marxist/Leninist theory. Factories, land, and crops were seized, but these measures only precipitated a breathtaking fall in industrial and agricultural production. Worker takeovers from factory owners and managers was simply licensed anarchy as the worker's/peasant's elemental mentality of stupid willfulness wrecked the country. For these folks it was a win/win situation; winning as one benefitted from the proceeds of selling a pilfered part of the factory, and winning again as one spiritually benefitted from the suffering consequent to the loss of work as the wrecked factory failed. Industrial production plummeted. Compared to the prewar year of 1913, industrial output in 1921 was down 70 percent, electricity down 75 percent, and sugar production down 96 percent. The grain harvests of 1920 and 21 were only 54 and 43 percent of the 1909-13 average. Two to five million died of starvation.

Things got so bad that Lenin, over Trotsky's early objections, persuaded the Communist leadership to reverse itself and temporarily encourage private enterprise. The program was called the New Economic Policy (NEP), which in its seven years of existence managed to bring much of Soviet Russia's productivity back to prewar levels.

Meanwhile, important changes were taking place in the Communist Party leadership. Stalin as Commissar of National Minorities was publicly rather obscure, but he was a careful and diligent worker building alliances behind the scenes. Trotsky, first as Commissar of Foreign Affairs, then as Commissar of War had perhaps been the single most influential person in saving the Bolshevik revolution from collapse. After defeating the Whites he turned his attention to the mounting economic problems where he was inclined to solve them with force. He was brilliant, but not as in tune with the subtleties of organizational infighting as Stalin. Lenin as chairman of the Council of People's Commissars and the spiritual father of the revolution ran the show. These three, with four others, formed the top policy making committee of the Communist Party, the Politburo. In 1922 Lenin appointed Stalin General Secretary of the Central Committee of the Communist Party in order, as the contemptuous Trotsky put it, to take care of the paper work. At the time it was a rather insignificant post. The paper work, however, included the power of appointment throughout the Party, a power Stalin was quick to use for his own ends. Accordingly, before long the post of General Secretary became preeminent in Soviet Russia and remained so till Gorbachev's fall.

Lenin suffered a series of strokes in 1922 and 1923 which made him an invalid till his death in January 1924. As he saw his ability to lead ebb away and observed the struggle for succession developing around him he prepared a "Testament" expressing his concerns about Stalin and suggesting that Stalin should be removed as General Secretary. The Testament was not seen by the Politburo until after his death. It was too late. The alliances Stalin had been making were too strong and the document was ignored. Thereafter, throughout the 1920s Stalin alternately made alliance within the Politburo with one faction then another, on each occasion weakening the opposition and strengthening himself. Trotsky, the obvious successor to Lenin, never had a chance. He was maneuvered aside, isolated, disgraced, expelled from the Party and exiled abroad where he was hounded from country to country till he was assassinated near Mexico City in 1940. Stalin's dictatorship, which began with the Fifteenth Party Congress in December 1927 would usher in an era of unprecedented barbarity where even his momentary allies of convenience in the Politburo would all be executed in the Great Purge a decade later.

With Joseph Stalin's rule came the resolution of a debate that had been nagging the Party for more than a decade; what was to be the scope of the Marxist revolution? Trotsky had envisioned it to be necessarily international with Russia leading the way, but this was obviously not going to happen, at least not for a long time. Stalin recognized that Soviet Russia was alone and very vulnerable in its Marxist revolution. He was therefore driven, not from deep theoretical grounds, but from pragmatic self preservation, to focus on "socialism in one country." To this end he now directed all of his energy and all of the energy and blood of the Russian people. The task was to make Soviet Russia an industrially mighty socialist state that was so strong no capitalist nation in the world could threaten it.

The time for petty capitalism was over. The country had more or less recovered from the birth trauma socialism; it was now time to put aside the temporary expedient of the NEP and start the real work.

The First Five-Year Plan began in 1928. Its goal was to start the transformation of Soviet Russia from a backward, agrarian nation into a world class industrial giant. The emphasis was to be on heavy industry. The planned annual growth rate in the range of 20 percent was extremely ambitious, but Stalin was in a hurry to catch-up, "We are fifty or a hundred years behind the advanced countries. We must make good this distance in ten years. Either we do so, or we shall go under."

The First Five-Year Plan also had another, less explicit objective, to cleanse the country of the capitalist infection allowed to develop during the NEP. As the NEP progressed the peasantry refused to sell its grain in a timely manner, hoping thus to drive up prices. The kulaks or somewhat better off peasants were seen as the main source of this problem. Indeed, the kulaks had always especially annoyed the Party because of their capitalist tendencies. Their independence was intolerable to the government. At the same time, the NEP had encouraged entrepreneurial Russians to become independent businessmen. They were often quite successful. However, their energy and entrepreneurial spirit were so intensely disliked by the average Russian as well as by the government that they acquired a pejorative name of their own, "Nepmen."

Kulaks and Nepmen were targeted even before the industrialization plan began. The Nepmen were taxed and forced into cooperatives and later their enterprises were declared criminal speculation. The kulaks, who in fact amounted to a sizable fraction of the peasantry, were marked for "liquidation of their class." When the peasantry withheld their grain Stalin had it confiscated. When Stalin was warned that wholesale confiscation would reduce seed stock for the next year and deprive the peasantry of food to survive till the next season he insisted still. The result was another famine in 1932 and 1933, but unlike the famine of 1920-1 which was caused unexpectedly by bungling and a bit of bad weather, this famine was planned by one man; Joseph Stalin. The government called it a "food shortage," but otherwise kept the magnitude of the disaster a state secret for a quarter of a century. We now know that at least 6 million died of starvation. The kulaks as a class were indeed liquidated, for those who didn't starve were often shot or exiled to Siberia. (One of those exiled for nine years was Andrei Gorbachev, Mikhail Gorbachev's grandfather.) The rest of the peasantry were eventually gathered together into large farm collectives.

The annual growth rate reached perhaps 10 to 14 percent in the beginning, and though it was well short of the goal, it was an exceptionally remarkable achievement. Without a doubt the intense push to industrialize was successful - at least superficially. The First Five-Year Plan was followed by two more before the onset of World War II. The plans were characterized not only by the novelty of a completely planned economy from above, but by the inevitable requirement of severe restrictions on worker freedom to move or change jobs, by an increased individual work burden, reduced standard of living, inefficient production, and very poor quality product. Soviet Russia was looking a lot like pre-emancipation Russia.

The necessary control of the people, as well as Stalin's paranoia, precipitated the Great Terror that began in 1934. No one was exempt. Purges between 1934 and 1938 reduced the Communist Party by 800,000, a third of its total membership. When state defined production norms weren't met it was often made the basis for charges of sabotage against any manager, technician, worker, or peasant who happened to be near. Exile, forced labor, or execution frequently followed. In 1937 the Red Army was purged with more than half of its senior officers shot. The so called show trials that began in 1936 led to the execution of four of Stalin's former Politburo colleagues. All of this was handled by Stalin's secret police.

Stalin's secret police resembled any number of its historic Russian predecessors, but perhaps most of all the *Oprichniki* of Ivan the Terrible. (Interestingly, Stalin was fascinated with Ivan the Terrible and resembled him even to their shared paranoid psychosis.) Within weeks of the Revolution in 1917 the Bolsheviks created the Cheka to, as they put it, destroy counterrevolutionaries. After several name changes reflecting its slightly changing mission it was called the Internal Affairs Commissariat (NKVD) in 1934. Eventually it would evolve, again with several name changes along the way, into the well known bogeyman of the cold war, the KGB. In the meantime, however, the first NKVD Commissar, Genrikh Yagoda, couldn't avoid becoming a victim of his own apparatus and was tried and shot in 1936. His successor, Nikolai Yezhov, the NKVD Commissar who did this for Stalin, was himself tried in 1939 and "disappeared." (The best known of the secret police leaders was Lavrenti Beria, a Georgian like Stalin. He continued to be a henchman of Stalin's from his first appointment in 1938 until Stalin's death, when shortly thereafter he was tried and shot. In a fascinating parallel Beria was professionally and physically remarkably similar to his contemporary in Nazi Germany, the SS and Gestapo chief, Heinrich Himmler.)

The pressure to industrialize steadily increased through the 1930s as Hitler became more powerful, but the emphasis shifted to war production. This caused a further hardship on the Russian people as even fewer consumer goods were available and the indirect benefits of heavy industry such as better transportation and more electric power declined. Aid was sought abroad, and many skilled technicians were brought in to make up for native shortages much as Peter the Great had done in 1698 and Alexander III in the 1880s. Nevertheless, sever shortages continued, and got worse in a kind of fairyland of unrealistic expectations. The usual explanation for the failure to meet planned goals was counterrevolutionary sabotage with even some of the imported foreign experts charged on occasion. The real cause was the often chaotic organization and supply, poor training and maintenance and a managerial corps disinclined to tell the truth in the face of possible charges of obstructionism and the consequent draconian penalties.

Stalin had hoped to forestall the inevitable by signing a nonaggression pact with Germany in August 1939. It had a secret sweetener for it agreed on spheres of interest where Soviet Russia got eastern Poland, Estonia, Latvia, Lithuania, and Moldavia (called Bessarabia at the time.) But given Hitler's attitude towards the Slavic peoples and his desire to find *lebensraum* (living space), and resources necessary for a growing Germany, it should have come as no surprise when Hitler attacked, non-aggression pact notwithstanding.

The attack came on June 22, 1941. And it appears that it was a complete surprise to Stalin, for he was reported to have been in a virtual state of shock for several days quite unable to make the decisions necessary to stop the German onslaught. On the other hand, this may have been an excuse to account for the embarrassing ease and speed of the German advance. In six months German forces advanced east 500 to 700 miles along a front extending 1,000 miles from the Baltic to the Black Sea.

The advance was no doubt facilitated by the cool regard many in the invaded lands had for their recent Soviet masters, but the scenario is too familiar to be explained by this alone. Over the centuries Russia had been deeply invaded many times, witness the

Mongols, the Lithuanians, the Poles, the Swedes under Charles XII, and the French under Napoleon to name only a few. Many of these invasions can be characterized by an early and humiliatingly easy defeat followed by eventual victory. As mentioned earlier geography and population have played their parts - yet this is still not enough to explain why Russians seem to take such beatings in their wars. One can't help but recall that suffering has historically been looked upon as efficacious, as necessary for salvation; the Russian soul seems to need suffering. But suffering has its limits, especially when a Russian feels an isolation from his Earth Mother Russia. Hence, invasion may satisfy certain needs, but if hostile to the Russian's sense of the inviolate Earth Mother (remember the thousand year old residual pagan "confessions to the earth" and the peasant's attachment to the land even as he was owned and abused by the master) then reaction and resistance follow. The Mongols largely let the peasant alone, and stayed in control for two and a half centuries.

After Stalin recovered his wits much of unoccupied Russia responded to the invasion with prodigious energy. A large part of Soviet industry was relocated to the Urals and beyond, out of German reach; factories were simply dismantled and shipped east to be reassembled and restored to full production in weeks. War production doubled and trebled. Workers increased their work hours and their productivity. In 1942 an unprecedented 55 percent of Soviet national income went into the war effort. Experience with five year planning now bore fruit, for it was only with such global planning from above that the economy could have worked at all under such a strain.

The combined civil and military Russian deaths in World War II have been estimated to be between 10 and 31 million. Even at the lower figure the devastation to the country was staggering, but behind the flash and smoke of destruction was a growing industrial establishment of enormous proportions that propelled the Soviets to superpower status by the war's end. Once again, the miracle was achieved through the indescribably hard labor and sacrifice of the Russian people.

The Russian people were sustained individually through the 1930s and 1940s by their will to survive and by an uncertain but important level of religious faith. Even Stalin after a generation of official atheism recognized the potential motivating power of the Russian Orthodox faith. When terror failed to move the Russian people to defend the country he turned to patriotism and religion; he called to the peasant's love of the land of Mother Russia. (Indeed, the self-focus was so strong that Russians have always called the war the "Great Patriotic War" only reluctantly acknowledging the contributions of their allies.) He used the Russian Orthodox Church to focus their religious feelings to inspire further resistance against the German invaders. This provided the social and cultural glue to hold the nation together just as it had in the past against the Mongols during the Tatar Yoke and the Poles during the Time of Troubles.

After more than 20 years of Communist persecution individual and institutional religious practice appeared to have substantially declined. Only a few percent of Russia's churches open in 1917 remained so in 1940. Figures vary widely, but it appears that somewhere between 20 to 50 percent of Soviet Russian citizens considered themselves believers as the war began. It is possible, however, that private religious belief and practice may have been higher. As in the past, the institutional church had lost its moral authority and only appealed to the people through the glories of the liturgy and the *starets*, its holy monks. Religion (as opposed to the Church) survived, but largely underground in what was called the "catacomb church." Further evidence of an underground church is reflected in frequent stories of secret baptisms. For example, in 1940 an NKVD official had his children secretly baptized, yet both he and his wife were Communist Party members, and secret baptism of children of army officers occurred as the Red Army occupied portions of Poland and the Baltic States where clergy was more available. It has been estimated that in 1970 half of all children born in European Russia were baptized, and it is alleged that even Stalin's daughter Svetlana Alliluyeva as an adult was secretly baptized in the Kremlin. An example of another kind of personal religion is the teacher who was also a Komsomol member and her husband who was the village police chief, yet they had an entire wall of their home covered with icons just as the 19th-century peasant home had its icon corner. When General Secretary Brezhnev was worrying about the United States and how it might use its nuclear weapons he once asked two religious leaders at a reception to pray for him. The remark may have been casual but nonetheless revealing.

As the resistance of the Russian people stiffened, the German army, with a little help from Hitler, made three fundamental errors; they treated the liberated peoples very poorly thus discouraging further defections and making Stalin look less odorous by comparison, they refused to make tactical withdrawals as circumstances suggested, and they hadn't prepared for a winter campaign. Like Napoleon and Charles XII this particular oversight cost the Germans dearly.

The year from mid 1941 to mid 1942 saw an immense territorial loss of more than half a million square miles with German gunners coming within sight of Moscow and Leningrad. But Russia's natural allies of territorial vastness and immense human resources began to prevail by the latter half of 1942, and in July 1943 she went on a front-wide offensive. With overwhelming superiority in men and weapons she recovered most of her losses by March 1944 in an impressive advance, though it is seldom noted that the rate was only about half that of the Germans going the other way two years earlier. Moreover, Soviet military deaths were twice those of the Germans on the Eastern Front.

The war's end in May 1945 left Soviet Russia and the United States as the two indisputable Super Powers. For Russia's part the power was essentially in the size of her military establishment and the possession of Eastern Europe as a buffer to the West. Her society was otherwise a wreck. Xenophobia quickly reasserted itself as she warily encountered the incredibly vital American culture with its equally threatening atomic bombs and leaders like General George S. Patton who openly advocated an attack on her. Western protests of peaceful intentions were unpersuasive. Stalin reacted; as Winston Churchill put it in his famous Fulton Missouri speech in 1946, "From Stettin in the Baltic to Trieste in the Adriatic an iron curtain has descended across the continent." Russia took refuge in isolation behind the Iron Curtain, and the Cold War began. Mutual fear and suspicion would drive an unbelievably costly arms race that would last until 1991.

In the meantime Stalin and his successors became more active in exporting communist ideology, most notably to China, but otherwise mostly to Third World countries. The initial Soviet successes terrified the West and particularly the United States which more or less blundered into a political plan of containment first annunciated by George F. Kennan in 1947. The political plan was given nuclear muscle, but Russia's own development of the bomb in 1949 made her substantially equal. Thereafter, for the most part, both nations played the game of geopolitics rather gingerly in the face of "mutual assured destruction."

To the utter relief of those around him as well as his more distant potential victims, Joseph Stalin died of natural causes on March 5, 1953. Surprisingly, his death was deeply mourned by many common Russians, for he had become a cult figure of virtual adoration by 1940 and was further glorified as the country's savior in World War II; it was all very sentimental, irrational, and submissive; and perfectly Russian. The figure in the mind of the average Russian was a totally fictitious construction built through the conscious efforts of Stalin and the unconscious force of a 1,000 years of history imbedded in the Russian psyche. He was looked upon not as the source of the people's woes but as the protector of the helpless, as "our little father," a phrase also frequently used in the 19th century by the peasants for the Tsars, and in the same warped context. Stalin facilitated this view by placing the responsibility for his own secret decisions and acts on his subordinates and then had them publicly tried and shot as circumstances required. Tsardom wasn't dead, it was only relabeled.

Chapter 11

The Fall of Soviet Russia

Affairs at home began to improve after World War II and especially after Stalin's death. After a period of political maneuvering Nikita Khrushchev became General Secretary in what was to be basically an eight year transition from Stalin's war Russia to an era of more or less "peaceful coexistence" and later "detente." Yes their were hitches along the way, notably the Cuban missile crisis which contributed to Khrushchev's eventual downfall, but Nikita was a man the West could live with. Unfortunately for him, his colleagues in the Politburo and the Central Committee weren't as flexible and he was retired in October 1964. His fate of comfortable obscurity instead of the firing squad dumbfounded the West so accustomed to tales of more draconian retirements.

Leonid Brezhnev emerged out of the Khrushchev ouster as General Secretary and lasted in the post for 18 years before passing on of old age in 1982. However, not only Brezhnev was aging in Russia during these years. Following Ibn Khaldūn's cycle of history the communist dynasty had peaked and was now in a period of decline, and so was the physical infrastructure. I have often described Russian advances through its history with qualifiers such as "apparently' etc.. The Brezhnev years would reveal another example of culture wide superficial advancement, for while Soviet Russia was competing everywhere in high technology, in space, in weapons, and elsewhere, her society was being stressed not strengthened by the experience. Modernity wasn't yet systemic in Russian culture. Add to this the growing privileges and isolation of the ruling class, the Nomenklatura, while the rest of the country stagnated, and one recognizes that eventually something would have to give.

Competition in an industrialized and high technology world requires an educated citizenry. One of communism's achievements was the transformation of a nation of illiterate peasants into 20th-century workers capable of meeting the challenge. By 1939 literacy had reached 81 percent, and under Khrushchev and Brezhnev it approached 100 percent due in part to training enough teachers to make possible classroom student -teacher ratios smaller than in the U.S. Science and technology were emphasized, and shortly the production of scientists and engineers surpassed that of the United States.

Soviet Russian science and technology were amazingly productive. In space they launched the first artificial satellite (Sputnik), then the first man in space, the first rocket to the Moon, and they were the first to circumnavigate the Moon and bring back pictures of its never before seen back side. In astronomy they built the Bolshoi Telescope, still the largest single mirror telescope in the world. In physics research their best theoretical physicists were comparable to the best anywhere. They built one of the world's most powerful particle accelerators ('atom smashers') at Serpukov for studying the fundamental structure of matter. Their development of nuclear weapons was only a heartbeat behind the U.S., and their nuclear reactor program was large and varied with power plants generating energy for the civil sector as well as for civil and military shipping. They built the Tu-104, the world's second jet airliner to enter commercial service (the first was the British Comet), and the Tu-144, the first supersonic passenger airliner.

But as is so often the case with Russian achievement, there is less here than meets the eye; Soviet Russian developments in technology have been a kind of technological Potemkin Village. Sputnik was a 184 pound shell with no useful payload. The scientific value of the Soviet Lunar firsts were marginal, largely due to inferior onboard instrumentation. The Bolshoi Telescope was scientifically useless because of poor planning and design, and the Serpukov accelerator never produced any important physics research. The Soviet nuclear power and weapons programs have been conducted in such a slipshod manner that the Russian people may be paying the price for radiation contamination for centuries (see Chapters 15 and 18.) The Tu-104 was so unreliable that it could not be profitably operated by foreign airlines, even when purchased at considerable discount. The supersonic Tu-144 was so problem plagued that it was permanently withdrawn from service. The curious Russian tendency of appearing not to be entirely at home with high technology appeared everywhere during the 1970s and 1980s. (Figure 6) On the other hand it is true that the Russians can produce one (or a limited number) of anything producible by the advanced nations of the world. Perhaps it's as the Soviet historian, Roy Medvedev has said, that when vitally necessary Russia puts *everything* into the product.

While Soviet Russia was making a show of its superficial technological achievements its industry and agriculture weren't doing so well. A low growth rate caused it to fall further behind the West. Typical of the problems was the acute shortage of computers. Russian technology was simply unable to support a meaningful computer industry, indeed, what she did have was almost totally dependent on technology transfer from the U.S.

The economy stagnated, helped in no small part by the burden of the arms race. By 1970 more than 60 percent of Soviet industry was supporting the military. Once again consumer products were either in short supply or of notoriously poor quality. In agriculture productivity was low because of poor weather, insufficient fertilizer, the indifferent attitude of the farm worker, and a transportation infrastructure that was so out of date a sizable fraction of the crop rotted in the fields before it could be shipped to market. In 1972 Russia began importing huge amounts of grain.

Consumer expectations weren't being met. Khrushchev had hoped to surpass the U.S. by 1970 and by 1980 reach the level of utopian communism envisioned by Marx. However, the standard of living which had been slowly rising through the 1960s failed to increase after the mid 1970s. Soviet economics wasn't up to it. It had failed, and by a big margin. In 1971 the U.S. produced 18 times the number of automobiles as Soviet Russia, four times as many trucks, and 600 times as many color television sets. Housing was no better with four to a room common. Khrushchev's 1956 optimism when he said that, "We will bury you" (economically) was never realized, and the Russian worker reverted to silent

Figure 6
A 1984 cartoon in the Soviet magazine Krokodil saterizes Soviet use of new manufacturing technology

suffering and his traditional suspicion of ambition, "We pretend to work, and they pretend to pay us."

However, not everyone was suffering. A new class developed in the "classless" Soviet society, the Nomenklatura. The Nomenklatura was a kind of 'old boys' network of perhaps 750,000 decision makers within the Soviet Union. One didn't have to be a member of the Communist Party to belong, though it was overwhelmingly probable that one was. The key was position, authority, and influence. If one had these, as a diplomat, KGB or Party official, scientist, or writer might have, then one might be on the secret list. Clearly this group was the elite; they ran the country. Indeed, they were the elite of the elite for they represented only about five percent of the Communist Party membership during the Brezhnev era which itself amounted to about six percent of the total population. To be a member of the Nomenklatura was to be a member of the privileged class in Soviet Russia. One had access to special stores where goods were available that most Soviet citizens couldn't even dream of, and while commoners were packed several to a room the new aristocracy enjoyed spacious and comfortable apartments and vacation *dachas*. The irony of Lenin's Marxism, indeed, its own "internal contradiction," was the irresistible seduction of power. The revolutionary vanguard got a taste of power and privilege and liked it. Thereafter, the Soviet Union was operated for the benefit of these few, and the rest remained part of the oppressed masses as if the Revolution had never occurred.

By the time of Brezhnev's death in November 1982 the Soviet leadership was thoroughly aware of the dangers of continued economic stagnation. The Politburo's election of the reform minded Yuri Andropov as General Secretary reflected the majority desire to do something about it. First steps were taken to control corruption, reduce the bloated and inefficient scientific and engineering community, and discourage alcohol consumption, but Andropov died of natural causes in February 1984, and his successor, Constantine Chernenko was content to sit out his dotage avoiding any change whatsoever. Chernenko passed quickly in March 1985.

When Mikhail Gorbachev became General Secretary it

immediately became evident that big changes were in store for the country. He was only 54 in a Politburo that averaged more than 70. He was young, dynamic, personable, well liked by the people, thoroughly reform minded, and willing to push for more than just superficial change.

His task was in some ways quite similar to that of Lenin's when coming to power in 1917; first he had to extricate his country from a hopelessly exhausting war, in this case the Cold War, and in particular the race to compete with the United States Strategic Defense Initiative, known perhaps more aptly as "Star Wars." There was just no way that Soviet Russia could successfully compete at this highest level of high technology weaponry. It was simply beyond her capacity technologically and above all, economically. Through vigorous diplomatic activity and daring unilateral 'tension reducing' acts Gorbachev largely succeeded. (He had some help with his greatest anxiety, Star Wars, for the American people were not thrilled with the staggering cost and technological risk either. Moreover, the arms race itself was getting economically out of hand under President Reagan.) Thus pressure was being brought on both sides to reach an accommodation.

Next Gorbachev had to deal with the inefficient and stagnant economy. Picking up on themes first considered by his mentor, Andropov, he revealed his plan for economic restructuring, *perestroika*. on a visit to Leningrad in May 1985. He wanted the economy to respond to market forces, to the needs of the people and not to fiat from above. He wanted individual factory managers to make the decisions about what and how much to produce and how to sell the product and for how much. Five year plans were bad, for they were much too rigid and insensitive to the real needs of the people.

However, his proposals were meeting with considerable resistance from the Nomenklatura. They were extremely reluctant to surrender any of their powers or prerogatives and were very disinclined to take risks, but this was just what Gorbachev was asking them to do. At each turn Gorbachev ran into their evasion and corruption. His solution, announced in mid 1986, was to bring the debate and reform process into the public domain by encouraging *glasnost*, meaning openness or frankness. He hoped the public spotlight would force the recalcitrant officials and managers to join the reform. It helped, but the grant of free speech also opened the door to far more and wide ranging discussions than he'd bargained for. Very quickly public debate began to chew over all aspects of the society, and for the first time in generations the very premises of Soviet life were being questioned. Among other things this led to open challenges of the idea of the Union itself, for the question of the absorbed nationalities had never been satisfactorily settled, particularly for the Baltic states, Byelorussia, and Ukraine. By 1990 matters were getting out of control as Lithuania declared its independence and the countries of Eastern Europe, which had been under Soviet hegemony since World War II, began to break away.

At this point Gorbachev's nerve failed him, possibly because he didn't have the stomach that circumstances required. (Lincoln, faced with a similar situation when the South seceded, sent in the army, and civil war ensued.) Daring in thought, cautious in action, aware of the radical novelty of his ideas and the economic reality requiring them, he recognized but underestimated the resistance of the conservative establishment. He also misjudged the willingness of the people to voluntarily bear the burden and uncertainties of the transition. In this milieu conservatives who thought he was going too fast gathered around Yegor Ligachev, while radicals who thought things should go faster sided with the opportunistic Boris Yeltsin. Gorbachev dealt with this by attempting a balancing act between them, but in the process he surrendered the initiative, the progress of his plan stalled, and public dissatisfaction grew. Before long his popularity had evaporated as public discontent with growing shortages mounted. Meanwhile, Yeltsin's star continued to rise through 1990 and 1991 as he focused on Gorbachev's failures.

By mid 1991 Gorbachev was clearly fighting a losing battle. He hoped to relieve the situation somewhat by granting considerable autonomy to the 15 Republics of the Union. The Union Treaty was to have been signed on August 20, 1991, but it never happened, for on August 19th an incredibly inept coup d'etat was attempted by an ad hoc, conservative cabal. Though it failed after three days, it did succeed in totally neutralizing Gorbachev as Yeltsin became something of a momentary national savior. Within two weeks the Baltics, Ukraine, Moldova, Uzbekistan, Kyrgyzstan, and Azerbaijan had unilaterally declared their independence and an interim central government had been set up whose sole purpose was to oversee the dissolution of the Union of Soviet Socialist Republics. The Super Power colossus had completely collapsed with hardly a whimper and the participation of only a couple of thousand of its citizens. The rest of the country looked on as passively as it always had on such occasions through the centuries.

On the evening of December 25, 1991 the red banner with its idealistic hammer and sickle was lowered for the last time from atop the Kremlin fortress. Within, Mikhail Gorbachev was making his resignation speech live on TV. In a poignant confession reflecting not only on his failure but also the failures of socialism in Russia, and incidently revealing a bit of Russia's self image, he said in part, "We had a lot of everything, land, oil, and gas; other natural resources. And there was intellect and talent in abundance. However, we were living much worse than people in the industrialized countries, and we were increasingly lagging behind them. The country was suffocating in the shackles of the bureaucratic command system. Doomed to cater to ideology, and suffer, and carry the onerous burden of the arms race, it found itself at the breaking point. All the half-hearted reforms, and there have been a lot of them, fell through one after another. This country was going nowhere, and we couldn't possibly live the way we did. We had to change everything drastically." The Soviet Union was no more. The socialist experiment failed completely.

The Union shattered into 15 independent republics and Boris Yeltsin became president of the historic core, the Russian Federation. The old Russian flag returned sans imperial eagles, with its broad white, blue and red horizontal stripes. Returning with it, if they ever left, were familiar problems that have haunted Russia for 1,000 years. The rate of alcoholism drove Yeltsin to ban its advertising in print and on TV in February 1995. (This is rather ironic since Yeltsin's own drinking problem is a matter of public spectacle.) Inflation took off, reaching as much as 25 percent a month in 1992, as agricultural output declined, and the gross domestic product fell three percent in 1990, nine percent in 1991, 19 percent in 1992, and 12 percent in 1993. Free enterprise is developing, but Russian entrepreneurs have still to fight the government as well as material shortages and organized crime. Generally however, they do not fight the Russian Mafia, they just pay 'protection' money. As the 20th century ends Russia is back where it was as the century began, a poor second to the West. In 1993 it had less than one-fifth the paved highways of the United States, less than one-fifth the number of telephones, and most significantly, much less than a thousandth of the electric production.

Yeltsin has had no more success in dealing with Russia's declining standard of living than Gorbachev, and not surprisingly his popularity has followed Gorbachev's. His clumsy handling of breakaway Chechnya and the revelation of the army's impotence have embarrassed him all around. Historically there is doubt that Boris Yeltsin is a transition figure anyway. The recent popularity and rise of General Alexander Lebed to be Yeltsin's National Security Adviser is more in harmony with Russia's past. Lebed is ambitious, wily, authoritarian, and appears to have a good sense of the historic needs of his people. Moreover, he now (1996) controls Russia's police and military.

What will follow is impossible to say, but all the misty talk of "democracy" in Russia flies in the face of its entire history. Russia will have its autocrat one way or another.

Chapter 12

Conclusion

We have seen ever so briefly 1,100 years of the Russian tide of history. The tidal metaphor implies both continuity and change. We see continuity in Russian passivity and submission beginning at least with Polyanian "meekness" and Viking rule described in the 12th-century *Russian Primary Chronicle*. We see the gradual surrender of peasant freedom from the 10th through the 18th centuries. The few rebellions that occurred seemed to object not so much to enserfment as 'unfair' enserfment. In this gradual enserfment we are witness to change even as the continuity continues. By the 19th century Western influences precipitated a relatively quick reversal of this trend as the intelligentsia began to define new rights of the common man. Tsardom was discredited and the peasant-turned-worker momentarily broke free of his old habit of submission and took up an experiment with Communism. Nevertheless, the habit remained strong, and the population reverted back to a posture of servitude, indeed, almost deifying their new master just as they had often done with their Tsars. At the same time a taste of self determination had been savored and not forgotten. Under further Western influence and the natural weakening of the Soviet power structure, the Russian people insisted again on their freedom in the last decade of the 20th century. We are witness now to this dynamic in the tide. The millennium habit of submission is being tested again. How it will turn this time is an open question, but it would be surprising if there weren't at least one more reversion to the old ways.

The role of religion and ethnicity has been evolving as well. Russians were, and are, Slavs before Orthodox Christians. In the 10th century Orthodoxy was consciously selected as the most suitable religion for their Slavic needs, and even then several centuries passed before conversion from their pagan origins predominated. Russian spiritual needs selected Eastern Orthodoxy and modified it over a half millennium to finally declare itself the Russian Orthodox successor to Rome and Constantinople, the Third Rome. The rulers gave the Church official status at the very beginning, so in just a few centuries it became largely an instrument of the state where political matters took precedence over spiritual. The Russian people, however, were instinctively very careful not to confuse official religion with their spirituality. If the Orthodox Church could supply a national coherence as it did during the Tatar Yoke, the Time of Troubles, and the Great Patriotic War, or provide succor and holy role models as the monasteries and elders did, or present an uplifting experience with its magnificent liturgy, then so much the better. Otherwise, what the Church wouldn't or couldn't provide was supplied by deep personal values springing from their Slavic origins and reinforced by worship in the home at the icon corner.

In recent times religious practice has been substantially impacted by the gradual urbanization of Russia. Under the force of communist industrialization urbanization exploded going from 15 percent in 1920 to 70 percent in 1990. With forced urbanization came not only a terrible economic burden for the future (cities were largely artificially constructed as centers of specialized production, thus requiring inordinate rail resources to supply them with goods other than their specialty,) but also a population wrenched from its rural, peasant mother culture and thrust into the sterile urban milieu. In the process many former Russian peasants appear to have given up their religion and turned to secular ideology, particularly Communism. Encouraged by the government's official atheism Russian religiosity was substantially weakened. But what of Russian spirituality and ethnic identity?

There are numerous reports of a very substantial return to the Russian Orthodox Church, although there is also some evidence that the old rewards of efficacious suffering are being rejected. Until the mid 19th century suffering was felt to a large extent to be desirable. This notion appears to be changing as Russia savors the delights of Western culture. There is a huge ambivalence here, however, for as the taste for Western ways develops, the old and cherished Slavic values pull in another direction. The Slavophils of the mid 19th century grappled with this problem and largely rejected the West. In this rejection we also hear echoes of Russia's suspicion of foreigners, historic isolation, self absorption, and messianism.

One hardly knows when the suspicion of foreigners began, perhaps under the Tatar Yoke. Certainly Russia has been invaded frequently enough to make "foreigner" equal to "invader." Moreover, foreigners viewed as 'other' are not of the chosen, the nation that will save the world through its spiritual purity. Therefore, foreigners are spiritually inferior and their approach to be guarded against lest they infect the true believer with heretical ideas. Communism's isolating paranoia has illustrated this innumerable times. In a typical recent example following the Chernobyl disaster a Soviet physician publicly thanked a volunteer American doctor for coming over to help, but wondered what his motive was.

Russian messianism has had a long history as well. We first saw it clearly with the Third Rome doctrine. It was equally clear 300 years later in 1833 with the Official Nationality doctrine, and with the Slavophil movement. Dostoyevsky later expressed it often and passionately in many of his writings, for example in 1877 in his monthly column, *The Diary of a Writer*, he wrote of Russia, "If it wishes to live long, every great people believes and must believe that in it, and it alone, is contained the salvation of the world..." A few decades later in an ironic, but perfectly consistent way, the writings of Trotsky and Lenin continued to reflect the notion of the special role of the Russian people in history, albeit without recourse to divine inspiration. Unfortunately, some might say, this Russian messianic conceit has facilitated continuing a value system that is not congenial to Western technology. The result has been habitual backwardness with only episodic and superficial relief. Russia today is in one of its periodic retreats from the exertions of its 20th century catchup episode.

At present the Russian historic tide somewhat resembles that of Islām; in each case great civilizations have evolved, for Islām the Abbāsid Caliphate among many, for Russia the early Kievan era. Thereafter circumstances eroded their greatness. The rise of the West with its high energy and alien values confronted each with almost irresistible cultural temptations. But the offerings of the West were also recognized as morally corrupting. For Muslims the answer remains undecided between a Muslim-Western synthesis or fundamentalism. For Russians the answer is also undecided; turn westward, effect some kind of synthesis, or largely reject the West and turn wholly Slavophil. It is fundamentally a search for identity.

The Russian problem can be illustrated by the way they have referred to themselves. In a political context they use *Rossiia*, the term probably coming into use sometime in the 17th century, possibly during the Time of Troubles. By contrast, Russians also recognized a different self, one implying a spiritual transcendence and specialness among peoples embodied in the term *Rus* and hence *Russkii*. To be *Russkii* was to be an Orthodox, believing son of Holy Russia. The specialness of the *Russkii* was in his soul, his spiritual dimension, especially as reflected collectively in Holy Russia and its historic, messianic mission.

To be *Rossiia* was to be a son of the state. When Peter III freed the nobility from compulsory state service in 1762 a new class was rapidly created, the intelligentsia. They were the only educated class at the time, and since they were now free of the state they were no longer *Rossiia*. However, because of their Westernization they were also no longer of the people, the *Russkii*. Thus the intelligentsia were never representative of the *Russkii* even though through their struggles to find their identity they explored and defined in great depth what it meant to be a *Russkii* and son of Holy Russia. This dynamic of the intelligentsia continued as the intelligentsia evolved into a middle class phenomenon in the 19th century. They desperately wanted to be *Russkii*, but they never really understood them. (Remember how the peasants turned in the young revolutionaries working among them, and the Bolshevik/Communist supercilious regard of the peasant.)

As industrialization and urbanization progressed the peasant himself was cut free of his roots, rather like the nobility experienced after 1762, thus weakening his traditional sense of being *Russkii*. Explicit religious belief declined to perhaps 50 percent or less of the population. However it was replaced by another messianic force and pseudo religion, Communism. The barrenness of Communism and its later leaders disenchanted the masses who experienced their subsequent suffering not as an efficacious vehicle to salvation through divinely sanctioned leaders, but as pointless suffering at the hands of evil men. The reaction was withdrawal into traditional Russian passivity which helped gain the country its notorious reputation for inefficiency and poor quality products. Most importantly, it has left Russians without a firm sense of identity, neither *Rossiia* nor *Russkii*.

The popular appeal beginning in 1992-1993 of Vladimir Zhirinovsky, who advocated attitudes and programs reminiscent of the early 19th century, and is looked upon by many as something of a neo-fascist, surely owes much to a longing to recover the national identity, to return to the Russia of old. No doubt part of the appeal of General Alexander Lebed has similar origins. And perhaps the need to be dominated and suffer is not gone either. It would be a perfect solution to Russia's dilemma if a pretender appeared claiming to be a descendant of the holy Alexander I. (Recall that his death in 1825 was questioned, many believing he secretly became a monk.) Or more plausibly yet, the appearance of a pretender claiming to be the offspring of either Alexis or Anastasia, the children of Nicholas II missing from the family's recently uncovered mass grave near the site of their execution in 1918. Otherwise, there are many legitimate Romanovs still alive, though some distance from the line of succession.

Russia has been a source of high art, science and technology, especially in the past 200 years. And in the last half of the 20th century her scientific and technological achievements have been second only to those of the United States. She has made important contributions from theoretical physics to materials science. She has led the way in space in many instances, has had a strong military technology, and a vigorous nuclear power industry. Yet, standing cheek-by-jowl with its innovativeness and modernity there continues to be a kind of familiar backwardness. Highly innovative mathematics and theoretical physics are done in a professional vacuum directly resulting from their self imposed isolation. In agriculture Russians thrust tractors into the fields then behave as if they've never seen a machine before. So much irrigation water is taken from the Syr and Amu rivers feeding the Aral Sea, once the forth largest inland body of water in the world, that it is literally vanishing. Heavy industry and nuclear power programs are carried out as if there are no secondary ramifications, for example the necessity of an adequate transportation network, and some attention to waste disposal. Environmental pollution, even by the lax standards of several decades ago, is today awesome (see Chapters 15 and 18.) However, perhaps we judge too harshly, for today Russia is in a state of transformation as profound as the 1917 revolution or that effected by Peter the Great. How it will turn out no one can say, but the tide of history strongly suggests some likely directions; a recoil into suspicious backwardness with the usual eventual race to once more catch up, or an energetic effort to break free of the old customs, hang on to what she has achieved and use modern science and technology to sustain her place in world history. In any event, the leadership will most likely be authoritarian.

PART III

The Revolutions of Science and Technology

At the close of the 19th century the worlds of the very large and the very small were thought to be fairly well defined and understood, and everything in between was gradually coming to be understood and managed as well. Indeed, some scientists had speculated that there might not be much left of importance to discover. The Victorian Age that was just ending was a tidy, well ordered time. Science and technology had made their appearance with suitable decorum and polite applause from the masses, and even a little wonder, the electric light, the telephone, the horseless carriage and all that, but things went on pretty much as before. A citizen of the West in 1690 would not have been very uncomfortable with the Western culture of 1890, science and technology notwithstanding. Nevertheless, the rush to the 20th century was gathering momentum with hints here and there within the basic sciences that many surprises were in store for the next century. We can now look back and say, surprises indeed, for our century might most succinctly be described as the century of science. This science and its concomitant technology have so transformed the West that a citizen of 1890 probably would be as helpless as a child in the world of 1990 where his every move would be with things and concepts unknown and undreamed of in the time of our great-grandfathers.

The scientific/technological revolutions of the 20th century have not only changed the way we do things but the way we see things as well. Earlier, the physical world was seen as thoroughly understandable and controllable. This crowning conceit of man would eventually lead to all kinds of mischief including the real possibility of destroying the world either quickly through nuclear war or gradually through subtle and profound insults to the global environment. Moreover, since the scientific enterprise is largely the product of Western culture it has also become the instrument of Western political and economic, indeed, cultural dominance throughout the world as illustrated in the preceding sections on the Islāmic Middle East and Russia.

Today the notion is just beginning to dawn that not only is the power of science and technology limited, but that the traditional quality of its knowledge is limited as well. In principle, some things are knowable and some things are not. We can never simultaneously know the exact position and momentum of an electron, nor can we ever see the entire universe, and that which is knowable in principle may not be knowable in practice; predicting the chaotic phenomenon of weather for example.

The discovery of the power and limits of science and technology is part of the tide we are now about to consider. Let us look at cosmology, electronics, nuclear science and technology, chaos and fractals, and space, five important areas that are part of, arguably, the most profound tide of all history;

Chapter 13

Cosmology

The night sky is an awesome mystery. Who among us hasn't looked upward into the vastness of the starry night and not felt the wonder of it. No doubt a similar wonder inspired primitive man's curiosity and thereby the first efforts to understand and explain what was seen. Surely the wonder invoked questions about being, the universe, and our place in it. Early observations were remarkable for their detail and accuracy as attested by the history of the ancient Egyptians and Chinese, the Babylonians, the builders of the Stonehenge, and the Maya. The cyclic movements of the Moon and the planets, the procession of the 'fixed' stars with the seasons, and unique events such as the appearance of comets were all recorded even if there was no true understanding of what they were. This is generally how matters stood, notwithstanding the contributions of Greek, Arab, and medieval European observers, until Copernicus, Tycho Brahe, Kepler, Galileo, and Newton began the transition from observation to understanding in the 16th and 17th centuries. In particular, Galileo's improvement of the telescope and his new observations provided the data and inspiration for a qualitative break with the past. The growing mathematical sophistication of the 18th and 19th centuries (also originating with Newton) as well as better telescopes allowed the Universe to be studied in considerable, albeit highly circumscribed, detail at the opening of the 20th century.

Imagine attending a hypothetical lecture, "Modern Astronomy," in 1898 at the University of Chicago's newly built Yerkes Observatory in Williams Bay, Wisconsin. Behind the speaker would be the recently installed world's largest telescope with its 40 inch lens. Professor George Ellery Hale, a leading astronomer of the time, would have told us of the new technique of astrophotography and how it was helping in the discovery and measurement of variable stars, the known number having trebled in the past three years to almost 1,000, of the discovery of a new moon of Jupiter six years ago and the discovery of another Saturnian moon just months ago, of the first photography of the spectacular solar flares, and the pictures of Mars which continued to suggest canals and thus the possibility of life beyond the Earth. He surely would have described the astronomical applications of spectroscopy where star light gathered by a telescope is passed through a prism and split into a spectrum of colors each characteristic of a particular chemical element. With this technique the composition of the Sun and stars were being discovered, facilitating a model of stellar evolution, quite erroneous, that stars continuously contract throughout their life and that they typically live about 100 million years. He also would have mentioned the puzzling "spiral nebula" and the speculation on whether they were local - part of our galaxy - or outside of it. His measuring stick, however, would have severely limited him for at that time distances of a 'mere' 50,000 light years were still unthinkably large. (A light year is the distance a light beam travels in one year going at 186,000 miles per second, ie., about 5.9 trillion miles. A light beam can go the distance around the earth in just over an eighth of a second.) Another puzzle was the failure to discover the predicted innermost planet, already named Vulcan, which theorists had calculated must exist to account for the imperfect motion of Mercury around the Sun. Earlier in the century similar observations and techniques had successfully predicted the existence and location of the planet Neptune in accounting for the irregular motions of Uranus. Yet another puzzle was the source of energy for the Sun; no one had any idea of the answer for that one.

Hale might have summed up our 1898 knowledge of the Universe and our place in it by reflecting that in successive stages over the past 400 years the center about which man existed had been placed first at the Earth, then at the Sun, and now finally in the Galaxy; he might also have plausibly continued that this Galaxy was it, there was no more. He would have been dead wrong.

As so often happens in science those untidy little odds and ends were to be significant clues to knowledge of a universe infinitely grander than anything imaginable in 1898. In just seven years the answer would be at hand to the puzzle of the Sun's energy source as the young Albert Einstein published his Special Theory of Relativity. Eleven years after that Einstein's General Theory of Relativity would go on to explain the peculiar motion of the planet Mercury in the absence of the hypothetical Vulcan. About the same time would come discoveries by others, including Niels Bohr, that would not only explain the fascinating regularity of spectra from the Sun and stars, but would also contribute, with Einstein's work, to our understanding of the life and death of stars and especially the birth and evolution of the Universe. So let us now take a brief look at Einstein's Relativity and the quantum physics of Bohr and others that have provided the foundations not only for cosmology, but for so much of 20th-century science and technology.

Einstein was attracted to aspects of a problem bothering many physicists at the turn of the century. The question was, what vibrated as light and radio waves passed through space? What was the carrying medium of these waves? This mysterious, never detected hypothetical stuff was given the name, "luminiferous aether." What ever it was it must have been very peculiar for it had no detectable effect on the Earth, the planets, stars, and light. It caused no drag to slow down material objects that passed through it, manifested no optical effects, nothing. Yet for historical technical reasons which we needn't belabor here physicists felt a need for an aether no matter how amazing its properties. Another property of the aether was that it be the stationary reference substance against which all motion in the universe be measured. However, this led to physical contradictions. It was not a satisfying situation, particularly since it violated the very highly regarded physical principle that with uniform motion of one system relative to another it is impossible to say which system is moving and which is at rest. One can only say that they are moving with respect to each other.

As Einstein considered the matter he was also aware of experiments and theory that showed that no matter how fast a source of light was moving with respect to any observer, the measured speed of the light was always the same for all observers. For example, suppose one stood on a beach at night, aimed a flashlight out to sea, and turned it on. Suppose someone off shore measured the speed of the light beam as it shot out from the flashlight towards them. The answer they would get for the light beam speed would be 186,000 miles per second. Now suppose a moment later a supersonic plane passed over headed straight at the offshore party. As the pilot passed over the beach flying at 10 miles a second he aimed his flashlight ahead and turned it on. Naively, one might expect the offshore observer to measure the speed of that light beam at 186,000 plus 10, or 186,010 miles per second, the sum of the speed of light from a stationary source plus the speed of the plane. But innumerable experiments have shown that no matter what the speed of the source (the flashlight) or the observer, the measured result will always be 186,000 miles per second. The measurements just didn't make sense!

Nature was saying that there were no special frames of reference where uniform motion was concerned, and that the speed of light was always the same regardless of one's uniform motion with respect to the light beam.

Einstein's genius was simply to accept these two experimental facts, no matter how crazy, and explore the consequences. So he started with just two postulates based on these observations; 1./ the laws of physics are the same for all experiments no matter what the unaccelerated motion of the experiment (this necessarily excluded experiments under the influence of the acceleration of gravity, however on earth the influence of gravity in these matters is very, very small), and 2./ the velocity of light is always the same no matter what the measurer's motion with respect to the light.

The consequence of these two postulates was the Special Theory of Relativity published in two papers in 1905 entitled, "On the Electrodynamics of Moving Bodies" and "Does the Inertia of a Body Depend Upon Its Energy-Content?" It was entitled "Special" relativity because it dealt only with the special case of unaccelerated, that is, uniform motion ("General" relativity would come later to deal with all motions including accelerated motions due to gravity.)

Many fascinating predictions were made by Special Relativity, all of which have been experimentally verified, but only one interests us here and that is the famous, $E = mc^2$. Einstein showed in his, "...Energy-Content" paper that energy and mass were interchangeable, that indeed, they were different forms of the same thing. He somewhat gingerly suggested this as an explanation for the source of the energy coming from radioactive substances just being discovered, "It is not impossible that with bodies whose energy-content is variable to a high degree (e.g. with radium salts) the theory may be successfully put to the test."

Eventually it was also recognized that $E=mc^2$ nicely accounted for the source of the Sun's energy. The process is identical to that causing the explosion of hydrogen bombs; the nuclei of two hydrogen atoms are fused together to form one helium nucleus which weighs less than the original two hydrogens. The mass difference escapes in the form of energy, the blinding incinerating flash of a thermonuclear weapon, or the bright warm sunshine of a summer day.

As exciting as the discovery of mass energy equivalence, $E = mc^2$ was, the capstone of Einstein's work and the theory suggesting more exotica in the heavens than any science fiction writer could have imagined was his General Theory of Relativity. Ever since his 1905 paper he was unsatisfied with the limitations of Special Relativity, after all, few things exist in this world in the absence of gravity. So with 11 years of further reflection and a lot of hard work, including learning a whole new system of mathematics, he published his General Theory in 1916. The General Theory of Relativity had no restrictions because of gravity or other accelerations, indeed, it was itself a theory of gravity completely superseding Newton's 200-year-old theory.

General Relativity transformed the notion of gravity as a force of attraction between two material bodies, to a geometric property of space. Matter causes space to be warped. As other matter (or a light beam) passes through the warped region it deviates from its normal path (the path it would have taken had the space been empty) and falls (or more accurately, follows a straight line in warped space that appears to us to be bent toward the attracting matter) into a kind of space dimple. This deviation from its normal path we experience as a gravitational attraction or force. The more matter packed into a region of space the more the warping, or the greater our perception of a gravitational force. (So a white dwarf star with a density one million times as great as the Sun's has a gravity comparably greater, hence the statements one sometimes sees about a teaspoon of material in such a star weighing 1,000 tons.) What General Relativity says is that space itself is more tightly warped there, and accordingly, matter and light more sharply diverted from normal, straight line paths of unwarped space. We shall find later that General Relativity also has a great deal to say about the birth, life, and structure of the universe as a whole.

One can guess immediately that the General Relativity theory of gravity explained the odd motion of the planet Mercury. Indeed it did. The planetary motion was measurably peculiar because Mercury was close enough to the Sun to be moving in a more warped region of space. The space was warped so it no longer had the properties of flat space described by Euclid's geometry. (For example, in Euclid's geometry the sum of the included angles in a triangle equals 180 degrees, and the circumference of a circle divided by its diameter is equal to pi, 3.14... but in warped space this is not usually true.) However, it was Euclid's geometry on which all the previous orbit calculations had been based. The unaccounted for observed error in Mercury's orbit was 42.84 seconds of arc per century. Newton's law of gravity does not predict any of this. Einstein's law of gravity predicts 42.91. Not bad!

The other advance in basic science that established the foundation upon which much of 20th-century cosmology is based is the development of quantum physics with its incredibly successful theories of the physical microworld. Quantum physics has evolved in large part out of observations that in the microworld matter and radiant energy interact in discrete units or quanta, not in a continuum.

Niels Bohr developed the first modern picture of the structure of the atom and its quantum behavior in 1913. This was a staggering break with the notions of the structure of matter up to this time. It was only in the preceding two decades that the existence of atoms was grudgingly accepted by most scientists, and even then the theoretical picture was of a classical nature, that is, a very small, solid sphere, characterized by an electric charge distinct for each element. However, the contradictions inherent in this classical picture were too acute to ignore and forced a rejection of the classical model, no matter how painful discarding the classical model might be.

Perhaps two matters were the most influential in making the classical picture untenable; the atom as then envisioned was patently unstable, it could not last even a trillionth of a second. However, this was obviously not the case since, apart from the minor exception of radioactive materials, undisturbed atomic matter existed quite comfortably in a stable form for as long as anyone cared to watch. Second, the radiation given off by very hot matter, the spectra or 'rainbow' from the sun as well as the elements in the laboratory, showed characteristics that could be explained only by the assumption that changes in energy of the atom could take place only in discrete amounts. The microscopic world behaved in a quantized way. There were only certain "allowed" energy levels of the atom. This was such a change from the previously viewed world of continuously variable factors that many of the older scientists of the early 20th century never did accept the new physics. On the other hand many others, and especially the younger scientists, readily accepted the simplicity and beauty of the quantum explanation so that on balance Niels Bohr's quantized atomic model was the consensus almost immediately. He was awarded the Nobel Prize for physics in 1922, nine years after his paper describing the atom and its electrons whose orbits could only be at specific distances (ie., energies) from the nucleus. Any change in the energy of the atom, for example if it was heated, or otherwise disturbed by light or x-rays or such, had to be in specific amounts. That is indeed exactly what is observed.

The history of quantum physics progressed into the "golden age" of the 1920s and 1930s where understanding of the quantum nature of the microscopic world at the atomic and sub atomic levels developed with astonishing rapidity. As an intellectual tour de force its success has been almost unparalleled in human history. Further advances, about which more momentarily, have led, for example, to the theoretical value of a particular property of the electron (its magnetic moment) of 1.00115965246 compared to the measured value of 1.001159652193; the difference is just 0.0000000003. This is equivalent to measuring the distance from Washington D.C. to San Francisco to the accuracy of the thickness of a sheet of heavy paper. Phenomena were explained and predicted that were totally beyond consideration in classical physical terms. All of transistor technology (based on what physicists call solid state physics), all of spectroscopy, all of atomic and nuclear physics, much of chemistry and molecular biology among others, and of particular interest for us, particle physics, are unthinkable without the concepts of quantum physics. (The only comparable achievement has been General Relativity with its similar ability to explain the heretofore unexplainable, and the great accuracy of its predictions. Indeed, recent astronomical observations have verified General Relativity with 1,000 times the precision of the magnetic moment prediction mentioned above.)

It is painful to bypass so much of the amazing history and concepts of 20th-century physics, especially of Werner Heisenberg's Uncertainty Principle which says that it is impossible to simultaneously know a particle's position and momentum with absolute accuracy, and Bohr's Complementarity Principle which reflects on the wave-particle duality of quantum level phenomena saying that for complete understanding they can and must be described in terms of both waves and particles. The Uncertainty and Complementarity Principles speak not only to physics, but to the fundamental limits of knowledge as well. However, since that is not our subject here we must pass on touching only the theory of elementary (so called) particle interactions.

In the 1930s and 40s atomic and nuclear physics had provided insight into the evolution of stars. Their energy consumption, and spectroscopic knowledge of their composition, allowed reconstruction of their histories and calculation of their ages. These details have given more reliable way points in establishing our understanding of the cosmos as a whole. However, it has been only in the past 20 years or so that cosmology (and cosmogony, the study of the beginning of the universe, which we here take the liberty to include under "cosmology") has come to significantly benefit from knowledge of the physics of the very small. In the 1970s evolving elementary particle theory became sophisticated enough to address questions concerning what happened at the very beginning of the evolution of the universe.

For some time the consensus among physicists has been that all action in the universe, at any scale, takes place through the operation of just four forces or interactions. Two are familiar to everyone; the electromagnetic force, manifest in both electric and magnetic phenomena and observed when a magnet is brought near iron, or when one sees the affect of static electricity when combing one's hair on a cold dry winter's day, or also when a battery is charged and discharged. Indeed, this battery example of the electromagnetic force in action points to its decisive roll in all of chemistry and hence, all of life. It is also the decisive force in classical mechanics; when one billiard ball bumps into another forcing it to careen across the pool table we are witnessing the interaction of the electromagnetic force between the atoms of the two balls, or more accurately, between the electrons in the atoms of the balls. Some of us have witnessed the same force in action when our car collides and bounces off another car.

The second familiar force is gravity. (For simplicity we revert to the old way of viewing gravity as a force rather than a geometric property of space.) Without the force of gravity we would have no weight. Gravity is what is manifest as weight. Since mass is the source of the gravitational force, the masses of our bodies and the earth exert a force of attraction on each other. For the author that force is 200 pounds. The only thing that keeps him from being pulled down to the center of the earth is the counter-force of the electromagnetic repulsion between the electrons in the atoms of his feet and the electrons in the atoms of the earth's materials. Here the two forces balance, and he doesn't sink into the ground.

The remaining two forces in nature are relatively unfamiliar to the layman but are at least as essential for life, and indeed existence, as the electromagnetic and gravitational forces. The weak force is manifest in certain kinds of radioactivity and the stability of certain elementary particles, for example, the neutron, crucial in nuclear reactors, will, through the weak force, spontaneously decay into an electron and a proton in about 15 minutes if left free outside of an atomic nucleus.

The strong force is the most esoteric of all. It is typically encountered in the atomic nucleus where it acts as a sort of glue that holds all the nuclear particles together. For example, the copper nucleus has 29 positively charged protons in it that would instantly fly apart because of the repulsive electromagnetic force of their like positive charges. However, they do not fly apart as evidenced, for example, by the permanence of copper pennies. At the very short distances found in the nucleus the attractive strong force overwhelms the repulsive electromagnetic force and holds everything together.

Separate mathematical models exist for each of these four forces, and they have worked rather well, but physicists have suspected since early in the century that a single model explaining all four might exist. This long sought model is sometimes referred to as the "Theory of Everything" or TOE. Einstein worked unsuccessfully for half his professional life at the less ambitious quest for a "unified field" theory uniting gravity and electromagnetism. Elsewhere, with better success, Sheldon Glashow, Abdus Salam, and Steven Weinberg discovered a model uniting electromagnetic and weak interactions into what is now called the electroweak interaction. They shared the 1979 Nobel Prize in physics for this achievement. Since then physicists have continued with some hopeful progress in the quest to further unite electroweak with the strong force in a "grand unified theory" (GUT). Gravity, alas, remains largely intractable.

However, there are good hints in both experiment and theory about what a GUT and possibly a TOE will probably look like. The characteristic of interest to us is the belief that as one goes higher in energy, the behavior of each of the four individual forces begins to converge. Accordingly, electromagnetic and weak forces unite and become indistinguishable at an energy of 100 billion electron volts (the electrons of the electric current passing through light bulbs in your home have an energy of about 120 electron volts.) Thereafter, above this energy, these two forces are identifiable only as the unified electroweak force. This unification has been verified by experiments conducted at the world's largest high energy particle accelerators at the CERN laboratory in Switzerland and at Fermilab in the USA. At a very much higher energy (a trillion trillion electron volts) the electroweak and strong forces are expected to unite into one as yet unnamed superforce. Finally, at a bewilderingly still higher energy (10 thousand trillion trillion electron volts) the superforce probably unites with gravity. These last two unifications have to be inferred because no experiment is able to duplicate the extremely high energy regimes in question. Indeed, there was only one time in the entire history of the universe that such energies existed; and that was before the first billionth of a trillionth of a trillionth of a second.

So how do we get from Hale's hypothetical lecture in 1898, where there is virtually no knowledge of Relativity, quantum physics, atoms, 'elementary' particles, the fundamental forces of nature, galaxies, and the Cosmos and its origin and evolution, to the present? We've already noted Einstein's discoveries of Special and General Relativity in 1905 and 1916. Einstein recognized immediately that General Relativity not only addressed the issue of local gravity but of the large scale structure of the universe as well. The connection was that with the huge masses of the stars and galaxies and the vast distances found in astronomy only the very long range force of gravity was influential. Over the next few years Willem de Sitter in Holland, Alexander Friedmann in Soviet Russia, and Georges Lemaître in Belgium solved Einstein's General Relativity equations for different cases representing plausible alternate structures for the universe. In 1917 Einstein chose a solution assuming the universe as a whole was static, ie., no expansion. In fact, his gravity equation as derived a year before implied an expanding universe, but he thought this couldn't be. So he added a term to the equation (known as the cosmological term) to make the answer give a static universe. Elsewhere, Friedmann in 1922 and Lemaître independently in 1927 proposed a curved universe that was expanding, a notion that at the time was quite unique; the general feeling was that the universe was static and everlasting. The implications of this model were that if the universe was expanding it must have had a beginning in time and in space (or more properly, time and space themselves must have had a beginning). Today we call this model the Big Bang theory of the origin of the universe.

Enter Edwin Hubble. From his graduate school days the American astronomer Edwin Hubble had been interested in spiral nebulae and the question of whether they were or were not part of our home galaxy, the Milky Way. Using the new 100-inch telescope completed in 1918 at the Mt. Wilson Observatory he showed that the Andromeda Nebula was in fact extra galactic, located about one million light years away. He first announced his discovery in a lecture at Oxford University in 1925. We now know on the basis of better measurements that the Andromeda galaxy is a spiral galaxy very much like our own but about 60 percent larger and located about 2,200,000 light years away. Hubble continued to gather distance data on nebulae until he had a few dozen that were extra galactic. Meanwhile Milton Humason and Vesto Slipher were getting spectroscopic data on these objects as well. Hubble noticed that the spectra of the galaxies were shifted to the red end of the spectrum in proportion to their distance. He interpreted this red shift as due to a velocity away from us. This is an example of a Doppler shift. Many are familiar with the equivalent effect for sound where we hear the pitch change as we approach or recede from sirens or bells. If the distance is getting smaller. ie., the source of the sound and the hearer are approaching each other, the pitch goes up (for light, the color would shift toward the blue). If the distance is getting larger, then the pitch goes down, which in the case of light results in a color shift toward the red. Hubble realized that a galactic recessional velocity proportional to distance reflected a general expansion of the universe. His discovery of the expanding universe is now acknowledged as the single most important discovery in 20th-century astronomy. It was only later that Hubble discovered that Einstein's General Relativity equations had already been solved by Friedmann and Lemaître for the expanding universe case. By 1930 the expanding universe model was largely accepted throughout the scientific community, indeed, Einstein himself had accepted it, ruefully rejecting the "cosmological term" he had misguidedly put in his theory with the remark that it was the greatest blunder of his life.

However, apart from its expansion, the question of the origin of the universe, whether the Big Bang or some other model, wouldn't be clinched until the 1960s. Only in the past 30 years has it been possible to reconstruct in any detail at all what probably took place at the beginning. What is the Big Bang model? Late 20th-century cosmological theory says that the universe, all of space, all of time, matter, energy - everything, probably began about 15 billion years ago with its spontaneous explosion into existence. At that instant the entire universe was contained in an infinitesimally small point that began expanding outward just as the fragments of a building are thrown outward at the instant of an explosion within. By extrapolating backward from today's observed expansion we get not only the time when it began but the temperatures, densities, and energies of this era as well. Using these data, numerous observations, and quantum physics and the rest of physical theory, scientists have been able to deduce the following as the most plausible reconstruction. From the first instant of the Big Bang to the Planck time, 10^{-43} seconds afterwards (that's 0.000,000,000,000,000,000,000,000,000,000,000,000,000,000,1 seconds) space and time were without coherent structure, and indeed, the very concepts of space and time are without clear meaning for that era. General Relativity and quantum physics are completely inadequate to describe what was happening. As the Planck Era ends the average energy of the contents of the universe dropped low enough for gravity to become a separate force leaving the still united strong-electroweak force. Thereafter the universe of space continued to expand and carry its contents along with it. At about 10^{-35} seconds the average energy was low enough for the strong force to separate from the electroweak. Also about that time the universe underwent a brief hyper-expansion phase, known as inflation. During the inflation era, which lasted till about 10^{-33} seconds, the accelerated expansion of the part of the universe that we can now see went from a hundredth of a trillionth of a trillionth (10^{-26}) of an inch in diameter to about four inches. Of course, other parts of the universe that are beyond our vision, because they are so far away, were expanding at an equal rate as well. The distant galaxies observed by Hubble to be rushing away from us are possibly artifacts of conditions at that time.

At about 10^{-11} seconds the average energy was low enough to permit the electroweak force to separate into the weak and electromagnetic forces. Each of these transitions where a unified force splits into daughter forces represent the beginning of an era where new particles can stably exist and where a new relationship between matter and energy is established. What results is a coherent, quantitative picture of the birth of the universe that is supported by observations today. All of the transitions described for the first hundred billionth of a second lead very naturally to the era when the universe was about one minute old and its energies low enough, that is, the universe had cooled enough, to allow various forms of hydrogen, helium, and lithium to form. The calculated quantities of these elements agree remarkably well with the quantities observed in the universe today. For example current physics and the Big Bang model predict about 75% of the universe to be hydrogen and about 25% helium; this is just what is observed. More remarkable yet is the quantitative confirmation of predicted abundances of micro traces of heavy hydrogen (H^2), lightweight helium (He^3), and lithium.

Nothing much happened from the time these nuclei were formed a few minutes after the Big Bang until the universe was about 300,000 years old. At this time the average energy of all matter and radiation dropped low enough for complete atoms to form. Atomic nuclei could now grab and hold onto passing electrons. (Before this, atoms were largely if not completely free of electrons, ie., they were highly ionized. The electrons themselves were rushing about completely free as well. The whole mess is called a plasma, the same thing one finds in experimental fusion reactors and within bolts of lightning.) When unionized atoms became dominant the universe suddenly became transparent. Radiation was no longer blocked or modified by interaction with the electrons in the cosmic plasma; it was set free to travel indefinitely through space and slowly cool as the eons of time passed.

Between 1948 and 1964 at least three groups of scientists independently predicted the existence of this remnant of the Big Bang. They predicted that by now this radiation should have cooled to a temperature of somewhere near 3 degrees above absolute zero and have a very characteristic energy distribution called a black body distribution.

In 1965, Arno Penzias and Robert Wilson were baffled by unexplained noise in the microwave receiver system they were trying to develop at their Bell Telephone laboratory. Eventually they determined that the noise was coming uniformly from all directions in space and had a characteristic temperature of about 3 degrees. Only later (in an episode reminiscent of the theoretical anticipation of Hubble's expanding universe 40 years before) did they hear that such radiation had been predicted as a kind of echo of the Big Bang. They were actually looking at the afterglow of the Big Bang, the considerably cooled 'light flash' that had been freely traveling through the universe for 15 billion years until caught by their instrument. If our eyes could see it as readily as visible light we would see a uniform glow everywhere in the sky, and this glow would be from the universe as it existed in its infancy just .002%

of its current age. Their discovery was decisive in establishing the Big Bang as the preferred model of creation and won for them the 1978 Nobel Prize for physics.

A moment ago we mentioned "inflation" and passed on, but it is a concept too important to treat so lightly. Although the Big Bang model answered a number of questions concerning the birth of the universe and nicely explained some of the current observational facts, it was by no means complete, and perhaps even fatally flawed. Why? The Big Bang model was unable to explain why galaxies exist, why space on a cosmological scale is so nearly flat - Euclidian - (remember, matter in space causes it to be warped), and why the universe is as uniform as it is in all directions and for as far as we can see. In 1980 Alan Guth, a young particle physics theorist, was intrigued by these problems facing his cosmologist colleagues. With a fresh point of view he proposed a very ingenious solution; he proposed that at about 10^{-35} second the expanding universe naturally evolved into a state that for the briefest instant transformed gravity into a very powerful repulsive force (in contrast to our normal experience of gravity as an attractive force.) The result was the instant of inflationary hyperexpansion already spoken of. This expansion had the effect of minimizing the cosmic consequences of a wide range of possible initial conditions in the early universe (ie., before inflation) so that no matter what one started out with, after inflation one ended with the uniformity and flatness we see today. Very nice trick. Guth's model isn't some wild fantasy, it's based on good physics. However, as initially developed, it had some problems and was subsequently modified by others in the early and mid 80s. Nevertheless, the consensus at the moment is that inflation in some form is probably on the right theoretical track.

As noted earlier, it was only with Hubble's work announced in 1925 that scientists began to fully grasp the true extent and structure of the universe. Very rapidly the nature of many nebulae were discovered to be "island universes," (a term coined by Immanuel Kant) galaxies often similar to our own Milky Way Galaxy and at staggering distances. Before long it became apparent that galaxies are the building blocks of the universe. Throughout the 1930s 40s and 50s many more were being discovered and their distances measured. Eventually sufficient data existed to identify a hierarchic structure of galaxy grouping. For example, the Milky Way Galaxy with Andromeda and a couple of dozen or so other, smaller galaxies form the Local Group some 2 or 3 million light years across. The Local Group is associated with other small galactic groups and somewhat larger galaxy clusters, as well as with the very rich, 1,000 galaxy Virgo Cluster 70 million light years distant in the direction of the constellation Virgo. Together they form the Local Supercluster containing more than 100 galactic clusters. Superclusters often contain 1,000 or more galaxy clusters and span several hundred million light years. In 1980s it was discovered that galaxies in clusters and superclusters form long filaments and shells surrounding dark voids in space extending several hundred million light years. As of 1994 the visible universe is thought to contain more than 100 billion galaxies each typically containing 100 billion stars (many galaxies contain much less than this and many others much more; some contain 10 trillion stars).

Just as the 1920s and 30s was a "golden age" for quantum physics the 1970s, 80s, and 90s is a "golden age" for cosmology (and astronomy in general) because of the number of important discoveries being made and the very substantial improvement in the understanding of all aspects of the birth and evolution of the universe and the systems it contains. It is particularly satisfying to find how much quantum physics, the physics of the very small, has facilitated this understanding. Indeed, without quantum physics there could be practically no understanding of cosmology. The unity and articulation of knowledge in this instance seems almost miraculous.

This understanding has been facilitated in part by the development of new instrumentation. Since the 1950s the space program has provided the capability to observe the universe from beyond the earth's obscuring atmosphere. The Hubble Space Telescope launched in 1990 and brought to peak performance in 1993 has vastly extended our vision and acuity into the farthest, earliest reaches of the universe. The COBE satellite detector has detected for the first time extremely slight variations in the 3-degree cosmic background radiation originating from the universe when it was 300,000-years old. The variations are interpreted as the first hint of the formation of galaxies a billion years later. The development of CCD electronic light detectors coupled to telescopes has made light image recording 10 times more sensitive than the old method of time exposures with photographic film. Earth-based optical telescopes themselves have evolved from the 40 inch Yerkes lens (1897) to the single 100 inch mirror Mt. Wilson instrument (1918) and the 200 inch single mirror at Mt. Palomar (1950) to the 400 inch multi-segment mirror Keck Telescope on Mauna Kea in Hawaii (1990). Elsewhere there are several comparable instruments either just completed or under construction. Radio telescopes with their huge parabolic dish antennae, some with diameters as large as 1,000 feet, 'see' the radio emissions of cosmic objects as effectively as optical telescopes see light and give a whole new picture of the sky. (Indeed, it was one of these instruments that first detected a quasar; quasars are the brightest, most distant objects known in the universe.) We shan't belabor the importance of new instrumentation beyond noting one more example, the computer. The complexity of events, forces, and theory places much of modern cosmology well beyond the computational reach of old style science. Fortunately, the computer substantially eases the computational challenge. It allows scientists to explore the consequences of mathematical theory from calculating the abundances of the basic elements in the universe to exploring new models of the formation and evolution of galaxies. We shall touch at greater length on the impact of this particular tool in the next chapter.

Today there is considerable confidence in the overall picture of the birth, expansion, and structure of the universe just outlined. However, by no means have all problems been solved. For example, the inflationary scenario still troubles some theorists. Ironically, the alternative some have been urging is a reconsideration of Einstein's so called blunder, the "cosmological term." It turns out that by suitable adjustments to this term one can solve a number of observational problems. The give and take of science in action will eventually settle the issue, meanwhile, we can enjoy watching the mystery unfold in the pages of magazines such as *Scientific American* and *Astronomy*. There remain a number of related questions including; how can the apparent age of the universe be less than the apparent age of some stars; what was the universe like in the Planck Era; and what holds galaxy clusters together. This last question springs from observation that there doesn't appear to be enough mass in clusters to provide the gravitation to hold them together. It's called the missing mass problem. Others more accurately call it the dark matter problem remarking that visible, luminous matter (stars) may represent only part of the cluster's mass with dark matter forming as much as 90 percent. Answers to some of these questions have been tentatively advanced, but a consensus has not yet been achieved in the scientific community.

However, the answer to the missing mass puzzle has greater consequences than simply explaining what holds galaxies together. General Relativity says that mass curves space. Accordingly, the total mass of the universe determines the total curvature of the universe, and the total gravitational pull against its expansion. As it turns out right now, the total amount of observed and inferred matter in the universe is not known well enough to decide whether it is finite or infinite in extent (somewhat more technically, whether the universe is closed or open) and whether the expansion will continue forever, or eventually stop and then collapse back into a condition similar to its beginning. This latter scenario has been graphically called "the big crunch."

We have sketched in the briefest terms the development and content of 20th-century cosmology; it now remains for us to consider how this knowledge has impacted history and helped to define the century.

It takes little reflection to recognize that for all the wonder invoked by the starry night overhead or the amazing discoveries and theories of cosmology, virtually no direct impact is discernable in human affairs. Discovery of the expansion of the universe or the definition of the moment of creation holds little promise for building better home appliances, stopping war, or facilitating the birth of nations. Clearly, the 20th-century advances in cosmology reflect a tide in intellectual rather than political or economic history. Viewed as such its importance becomes immediately evident.

The smugness of the 19th-century Western mind has already been alluded to and stands in bold contrast with our views today. In addition to the terrible blow inflicted by World War I, the discoveries in physics and cosmology have effected the final *coup de grâce* to the West's notions of man's special place in creation. Relativity has declared that there is no special position from which man can make his measurements, that all reference frames are equivalent and equally valid, and that none is preferred. Quantum mechanics has said that our knowledge of the world, at least at the microscopic level, is limited and that the very act of observation will unpredictably change that world. Cosmology has shown us that our home galaxy is a rather common type in no way distinctive, and that there is no special place in the universe that could be called the origin. Cosmology suggests that the universe we occupy may not even be special, that the part out to 10 or 15 billion light years that we now see may be only a small part of 'our' universe and that other, totally inaccessible parallel universes may exist created simultaneously with ours at the moment of inflation. Cosmology also raises the possibility of cyclic universes, that the universe expands to some limit, then contracts to a point, then bounces back with a new big bang and the cycle repeats over and over again for ever. So we may not even exist in a special time.

As the philosophical lessons of cosmology and the other sciences disperse into the global popular consciousness through the next century our time will be perceived as the source of a new and powerful tide in history. Cosmology will be the example par excellence of the intellectual odyssey at its finest and a fitting complement to other, more direct scientific/technological achievements characterizing the 20th century. Perhaps the most influential of these is electronics, to which we now turn.

Chapter 14

Electronics

"He found electricity a curiosity and left it a science." So wrote Carl Van Doren of Benjamin Franklin's discoveries in the late 1740s. We are all familiar with the story of Franklin's kite experiment where he showed that lightning was electricity. His invention of the lightning rod is equally well known. But Benjamin Franklin's seminal contributions to the science of electricity are much broader than that as indicated by the terms which he introduced and which are still in common use today; armature, battery, charged, conductor, discharge, electrical shock, positive and negative, to name only a few. The development of this vocabulary and the underlying understanding it reflected clearly mark the beginning of an era of science and technology that has only in the 20th century burst forth with the overpowering influence we are about to chronicle.

Out of his experiments and discoveries Franklin in early 1749 was, "Chagrined a little that we have been hitherto able to produce nothing in the way of use to mankind." Nevertheless, by the end of the year he had invented and recognized the practical importance of the lightning rod and later used an electric shock to kill his Christmas turkey in 1750.

The evolution of the science of electricity into electronic technology was significantly facilitated by the researches of Charles Coulomb (1736-1806), Alessandro Volta (1745-1827), Andre Ampere (1775-1836), Georg Ohm (1787-1854), Michael Faraday (1791-1867), and Joseph Henry (1797-1878). Some readers may recognize these names assigned to the units of electrical measure for charge, electric force, current, resistance, capacitance, and inductance. Most notable, however, were the contributions of James Clerk Maxwell who in 1864 developed a series of equations showing the essential link between electric and magnetic phenomena and the way in which these were propagated through space. Twenty-two years later Henrich Hertz experimentally verified the existence of Maxwell's predicted radio waves, and just nine years after that the young Guglielmo Marconi invented wireless communications. Radio, crude as it was, was born just at the threshold of the 20th century. Meanwhile Thomas Edison was finding innumerably more direct uses for electricity from the electric light to improved telegraphy.

One of the earliest applications of what today we would call electronics was Marconi's wireless radio. Indeed, the first recorded use of "radio" in this meaning occurred in 1903. In this connection it is also interesting to note that the word "electronics" is not even listed in the 1933 Supplement to the *Oxford English Dictionary;* it was listed for the first time in the 1972 Supplement. There can be no doubt that electronics as we know it took its early impetus from the development of radio. The desire to improve radio communications inspired the invention of the crystal, or "catswhisker" receiver patented in 1906 after a fellow named Greenleaf Whittier Pickard discovered that a fine sewing needle placed across carbon blocks would detect radio signals. No amplification of the radio signal was achieved by this device, it merely processed the signal so that the audio portion could be heard in a pair of headphones. In 1904 J.A. Fleming, pursuing earlier observations by Edison, invented the vacuum tube diode, which later became an improved substitute for the catswhisker/crystal assembly. Lee DeForest in 1906 added a third element and invented the vacuum tube triode thus making possible for the first time the electronic amplification of radio signals or any other type of electrical signal. The vacuum tube reigned supreme in this function for the next 50 years.

The telephone was also a major instrument of communications. Its development was largely in the hands of the Bell Telephone Company which operated a very respectable system of laboratories. Respectable indeed, Bell laboratories have facilitated the research of no less than seven Nobel Prize winners in the past 40 years! Three of the winners produced what is arguably the single most influential invention in recorded history, the transistor.

After World War II the Bell Telephone Company recognized the limitations of their mechanical switching systems in the face of the enormously growing use of the telephone. Call connection between sender and receiver was just too slow, expensive, and unreliable using mechanical relays and human operators cross connecting wires at massive switchboards. Vacuum tube technology, now well established in the consumer market place and developed almost to perfection by the demands of war, nevertheless had itself some very serious limitations when required to perform on the scale anticipated by the telephone industry. Millions of tubes would be needed to perform all of the call switching operations. Unfortunately, tubes use a lot of electric power, most of which ends up as waste heat that has to be dealt with, often with expensive cooling or air conditioning. Moreover, the mean time between failure - the reliability - of tubes was acceptable for the relatively small scale use in consumer and military systems but entirely inadequate for installations using tens of thousands as in a typical telephone exchange.

With understandable corporate encouragement a group of scientists at the Bell Labs in Murry Hill, New Jersey were wondering about semiconductors as a possible alternative to tubes. At this time not much more was known about semiconductors and what they could do than was known a half century earlier with the use of the galena "catswhisker" crystal detector. A semiconductor is a metal of intermediate electrical conductivity between a good insulator and a good conductor. Silicon and germanium are typical examples. The conductivity of a semiconductor can be controlled over a very wide range. This control can be achieved both by introducing trace amounts of specific impurities as the material is being fabricated into a useful device, or by establishing suitable electrical conditions at the interface with other materials, as for example happens at the point where the plated copper "catswhisker" contacts the galena crystal. There was some hope that if semiconductor crystals could perform the function of Edison's and Fleming's vacuum tube diode, as they did with the "catswhisker" detector in early radio receivers, then perhaps they could be modified to behave as vacuum tube triodes and function as effective telephone switches. William Shockley had his first insight into this possibility in late 1939. The war and a lack of understanding of the underlying physics delayed eventual success until late 1947. The official announcement came in June 1948 with remarkably little notice by the technical and financial communities.

William Shockley, John Bardeen, and Walter Brattain had invented the transistor. Its superiority over the vacuum tube was patently obvious. The tube was bulky, relatively expensive, relatively delicate, not unusually reliable, and a very inefficient user of electric power thus requiring careful attention to cooling while also demanding expensive high voltage power supplies. Even the earliest transistor was hundreds of times smaller (about the size of a pencil eraser), and with features natural to its design gave promise of being extremely inexpensive to produce. It was also naturally a physically very robust device of very high reliability, efficiency, and extremely low power consumption, operating very nicely at low, easily produced voltages. Indeed, it is difficult to imagine a more perfectly suited device for control of electrical signals than the transistor, and this clearly accounts for its ubiquitous application today. In 1956 Shockley, Bardeen, and Brattain were awarded the Nobel Prize in physics for their achievement.

The transistor's phenomenal success was only just beginning with its application as a discrete component wired into circuits much like vacuum tubes had been. The manufacturing process for individual transistors is a natural for simultaneously producing a bunch of transistors already connected in a complex functional array, an integrated circuit. Transistors are made by depositing layers of various materials on a crystal of silicon and then etching through these layers in a pattern and sequence appropriate to the transistor's design. The pattern to be etched is projected on the surface of the deposited layer and photographically fixed there. Suitable solvents then wash away the photographically unexposed portions of the 'picture' of the transistor's structure leaving a clear surface to be etched. The negative containing the etching pattern can be of any convenient size. After it has been prepared it is put in a projector where the image is optically reduced and projected on a thin slice of a single silicon crystal called a wafer. Since this is a simple photographic process, very similar to the way snap shots are developed, it is easy to project any number of transistor 'pictures' simultaneously with a multi-picture negative called a mask and completely fill the wafer surface with transistor images. Hundreds of transistors can be made in one shot this way. After the etching and other processing, the wafer is cut up and each little transistor chip or die individually encased in its container with suitable electrical leads sticking out to be soldered into the final circuit.

The potential for production economy in this process is clear. Wafer size, reflecting the industry's ability to produce and effectively process large single crystals of silicon, has grown from one inch in diameter in the 1950s to six inches in early 1990s. By the mid 1990s eight-inch diameter wafers may become the standard, and some manufacturers are considering the possibility of going to 12-inch wafers by the end of the decade. Larger wafers mean more semiconductors per single processing pass.

This is also where significant potential appears for integrating a number of transistors into a completely predesigned circuit that is etched and produced as a single unit, the integrated circuit (IC). In July 1958 Jack Kilby at Texas Instruments got the idea of creating all the components of a complete circuit on a single silicon chip. The intention was to reduce the growing number of wiring interconnections that were beginning to plague the developers of the increasingly complex circuits made possible by the transistor. The problem was growing so acute that each of the military services as well as several semiconductor manufacturers were searching for a solution. Robert Noyce at Fairchild Semiconductor independently got a very similar idea for creating integrated circuits in January of 1959. Today Kilby and Noyce are universally acknowledged as the co-inventors of the integrated circuit.

At first, integrated circuits were so expensive that no commercial user was interested. In 1961, for example, they typically cost $120 each. On the other hand, the military with their interest in lightweight ICBM guidance systems, and NASA with the ambitious moon landing program getting into gear, were both ready and willing to bear the cost of integrated circuits. Indeed, ICs were sold exclusively to the government until 1964 when commercial utilization became cost effective. With this early government support the costs of learning how to make them were quickly recovered and production began to expand at a mind boggling rate. In 1963 five hundred thousand ICs were sold. Production was increasing every year so that by 1966 thirty-two million were sold. Of course this rate of increase couldn't be sustained for long, nevertheless, by the 1990s production of ICs surpassed one billion units per year. The natural effect of this success was to bring the price down - a lot. The chip that cost $120 in 1961 was selling for $1.27 ten years later and substantially under a dollar today for a similar function.

For a decade ICs were designed almost entirely as functional elements for larger systems. However, as time passed it became clear that it was possible to create complete systems on a chip. Thus was born in 1970 the computer on a chip, the microprocessor.

In order to get a better sense of this evolutionary avalanche from tubes to transistors to ICs to microprocessors we must go back and discover another current of our story, the development of the electronic digital computer.

The electronic digital computer is a system that receives data, transforms the data into a usable form, performs programmed operations on the data, and then provides an output. The inputs and outputs most frequently encountered are "alphanumeric" (letters and numbers) via a keyboard and a display device such as a video monitor or a printer. Sometimes the input is taken directly as an electrical signal from a sensing device of one kind or another, for example from a thermometer, or wind speed gauge, or turnstile counter, or a bar code reader. The output may also be an electrical control signal, for example that which controls the speed of one's car when set on "cruise control." However, a "digital" computer must have all input data transformed into digital form so that the computer can operate on it. "Digital" manipulation is chosen because it is invariably more accurate and often faster than "analog" where the signal is continuously variable, but difficult to define precisely. We see these properties in the differences between digital and analog clocks.

The digital form of data also perfectly lends itself to the operation of a switch or relay; the switch is either open (digit "0") or closed (digit "1"). Or in the case of a vacuum tube or transistor, the device is either turned off (digit "0") or turned on (digit "1"). It turns out that well before the invention of the electronic digital computer there already existed a system of arithmetic and logic based on this "binary" (0/1) digital way of doing things. It's called Boolean algebra. Accordingly, a computer's switches (relays, tubes, or transistors) have merely to switch on or off indicating the binary digits "0" or "1" in certain well prescribed ways to arithmetically or otherwise logically manipulate the data.

The data are transformed by the input device from the familiar numbers, and letters of the alphabet, into binary signals that can be used by the computer's central processing unit (CPU). The CPU then goes to work guided by a set of instructions called a "program" which is stored in the computer's memory. The memory also stores any intermediate results obtained during lengthy computations.

Writing the program in the necessary binary machine language of 1s and 0s is far too tedious and abstract for most of us and certainly is no way to encourage the common use of computers. This problem was solved with the introduction of what are called high level programming languages like Fortran, Basic, Pascal, and "C". Computer programmers spend their time writing machine language, and intermediate and high level programming language programs. Many of these make the computer more "user friendly" for the non-expert. Writing the software, the programs that tell the computer what to do, has become a major factor in determining the computer's continued success. There exist commercially available software to do just about anything from word processing, sophisticated graphics, and accounting to data organization and retrieval, scientific calculating, and entertainment.

The first notions of an electronic digital computer appeared shortly before World War II. Among others, Howard Aiken, faced with the prospect of doing lengthy scientific calculations by hand, was inspired to think about the problem and came up with the idea of an electronic computer defined in his 1937 memo, "Proposed Automatic Calculating Machine." After some difficulty in interesting the academic and business communities, a small business machine company, International Business Machines, decided to support his work. Aiken developed the Automatic Sequence Controlled Calculator in 1939 for IBM at Harvard. By 1943 Aiken's MARK I was supporting the war effort computing tables to direct artillery fire. The MARK I was a huge machine, 51-feet long and 8-feet high, using only electromechanical relays for its binary switching operations. It could do three additions per second. It did have the advantage, however, that slow as it was, it was very much faster than a human, and it never made a mistake. Indeed, its recomputation of some already existing artillery tables revealed numerous human errors. The military advantage this gave through more accurate artillery fire is obvious.

Meanwhile, the government in Great Britain had also recognized the power of the electronic digital computer, especially in British intelligence where breaking the German code was becoming a fruitful pastime. The Brits developed a special purpose code breaking computer in 1943 named Colossus that was in some ways more advanced than the MARK I then at Harvard. This machine used not only relay switches but vacuum tubes as well.

As World War II progressed the military and their governments were firmly persuaded of the benefits of computers and were to provide almost all the R&D cash and become virtually the sole customers for computers for several years after the war. However, others had an even broader vision of the computer; in 1944 *Popular Mechanics Magazine* wrote with great prescience of the MARK I and electronic computers in general that, "...it will accelerate the exploration of nearly every field of knowledge..." With hindsight this turned out to be a very conservative statement.

By the end of 1945 computers like the MARK I and Colossus had evolved into even more massive and faster machines. The ENIAC completed in November 1945 had 18,000 tubes, weighed 30 tons, and used 174,000 watts of power, but it could do a multiplication in three-thousandths of a second. The same multiplication took the MARK I six seconds. However, the transition from mechanical relays to vacuum tubes had its price; with so many tubes the system mean time between failure was a matter of hours.

In 1951 the first UNIVAC I was delivered to the US Census Bureau. It was not only the first commercially available electronic digital computer, but also among the first computers designed to reflect new organizational and operating principles defined in 1946 by the mathematician, John von Neumann, at the Institute for Advanced Studies at Princeton. Von Neumann's seminal paper, co-authored with two others, "Preliminary Discussion of the Logical Design of An Electronic Computing Instrument" has influenced the design of all digital computers ever since, and has been described in the Encyclopaedia Britannica as, "...a landmark in the history of computer sciences, if not in the history of human thought."

Computer design proceeded through the 1950s and 1960s with massive government support that was decisive in its rapid development. The benefit of decades of government funding is evident in every computer now made. For example, the development of computer memories, and especially magnetic core memories that dominated the field until the early 1970s was a direct result of government support. The same is true of printed circuit boards. There are virtually no wires in modern computers; everything is assembled on printed circuit boards, some with half a dozen layers or more. A printed circuit board is made of an insulating substrate with a copper coating. Through suitable photographic and etching processes reminiscent of the production of integrated circuits the 'wire' layout is defined on the board and the parts simply soldered on at the proper locations. The soldering process itself has been automated using a wave of molten solder that quickly washes over the board establishing the soldered connections without burning the parts.

The needs of the National Security Agency to decipher foreign codes spawned the Harvest computer complex with its Stretch processor that was thought to be the most powerful computer of the era from 1962 to 1976. The US Air Force's Air Defense Command has driven the development of several state of the art computer systems over the decades for use in the SAGE bomber defense system of the 1950s, the BMEWS intercontinental ballistic missile defense system of the 1960s and 70s, and the SDI "Star Wars" missile defense system of the 1980s. Nuclear weapons research contributed to the demand for development of the supercomputer with its prodigious computational power. The first CDC 6600 supercomputer went to the Lawrence Livermore

National Laboratory in 1966; the first CDC 7600 went there in 1969, and the first CDC Star 100 to the same place in 1974. These computers were also used by the US Weather Service for long range weather forecasting. And finally, in 1976 the first of the legendary CRAY supercomputers went into operation.

These machines represented the cream of the crop, the most sophisticated, most powerful, and most expensive computers available. In 1954 IBM's model 701 rented for $20,000 per month. The purchase price of more recent models could very easily exceed tens of millions of dollars, but speed went from a few hundred operations per second to millions, and data storage from hundreds to billions of bits of data.

However, the computer was too generally useful to remain the high priced property of the government. As business and scientific applications became sufficiently important the commercial market surpassed the government market in the early 1970s. Part of this was due to the introduction of computers less powerful than the room-size "mainframe" variety described so far. The table-top "minicomputer" came on the scene in 1965 with Digital Equipment Corporation's PDP-8. It cost only $18,000 and sold like hot cakes. More than 50,000 were produced. Five years later the faster, more powerful, and cheaper ($10,800) PDP-11 appeared eventually leading to sales of 250,000 units. In 1977 DEC came out with the VAX line of minicomputers that has set the standard for this type ever since.

The minicomputer, however, was not where the true revolution in computer design and utility lay. While the 1970s saw the minicomputer makers, like their mainframe colleagues, push their designs further and further, they were also becoming less sensitive to new ideas outside of their particular conception of a computer. These new ideas were embodied in the development of the integrated circuit and its inevitable extension, the microprocessor "computer on a chip", which returns us to the mainstream of our story.

We have seen how the development of the integrated circuit facilitated the incorporation of many transistors and supporting components onto a single silicon chip. The number of transistors that could be put on a chip was rapidly increasing as experience was gained with the integration process. In 1970 Intel's 4004 microprocessor chip had 2,000 transistors on it. My 'ancient' personal computer's 80386 microprocessor chip, dating from 1985, has 400,000 transistors on it. The Pentium chip from Intel has more than 3 million transistors, and in 1996 Intel expects to have available its "P6" microprocessor chip with more than 5 million transistors on it. Semiconductor manufacturers are confident that a memory chip will be produced by the year 2000 with one billion transistors on it. Indeed, the single chip microprocessor has become so powerful and inexpensive that Intel plans to combine 9,000 individual P6 chips to create a single parallel processor super computer capable of 1.8 trillion operation per second. It will also have 262 billion bytes of memory. (A byte is composed of eight bits. One bit is represented by the digital 0/1 choice.)

Feature size, which directly determines how small a transistor can be made and thus how many can be put in a given area, was reduced from six microns in 1970 to significantly less than a half micron today. By way of comparison, a human hair is about 50 microns in diameter.

Chip size is partly determined by how many chips can be made from a single wafer and thus is partly dictated by the economics of production. The first microprocessor chip in 1970 was about an eighth of an inch square. The Intel Pentium chip of 1993 is about three-quarters of an inch square. As we have seen, wafer size has grown from one inch diameter to six to eight inches. This has tended to keep the number of chips per wafer relatively constant and thus production efficiency high. Also, as the size of a chip increases there is a greater probability that dust in the air will settle on it and damage the circuit during production. This is the single most important source of chip loss. Accordingly, the development of extremely clean production environments has had a decisive impact on the production yield and final price of ICs.

The high computational power and low price of microprocessor ICs has historically been irresistible. From the very beginning in the early 70s they were used in scientific and industrial instrumentation and shortly thereafter incorporated into home appliances and computer kits for hobbyists.

This seemingly innocuous amateur activity of putting together small home computer kits to sell to friends and in local hobby stores was the beginning of an incalculably important industry. In 1974 a few hobby systems were designed and a couple of thousand kits sold. These were literally garage enterprises. Within two years there were enough kits available to fuel the needs of a surprisingly large number of interested hobbyists. These hobbyists frequently formed computer clubs so that by 1977 there were hundreds nationwide with five in the San Francisco area alone. One of these was the "Home Brew Computer Club" with several hundred members, two of whom were Steven Jobs and Stephen Wozniak.

The enthusiasm for the first kits appearing in 1974 and 75 was so manifest that Jobs and Wozniak decided to get in on the action, produce their own kit, and make some money as well. Their Apple Computer Company was formed on April 1, 1976, and its first product was simply a printed circuit board with IC computer chips soldered on it. The Apple I had no keyboard, no monitor, no power supply, not even a box to house it, but it sold modestly well in the San Francisco area for the price of $666.66. Encouraged by this success they embarked on the design of the Apple II which was eventually housed in the familiar beige case that has marked almost all personal computers ever since. It also had a built in keyboard and the ability to present crude color graphics on a TV set. By early 1977 they had obtained sufficient financing to go into serious production, and the rest is indeed history. By 1980 sales reached $120 million, and $600 million two years later. The age of the personal computer, the PC, had arrived.

When IBM introduced its first PC in August 1981, it essentially split the market into two incompatible areas, those that used software designed for Apple computers and those that used software designed for IBM machines and their clones. The IC microprocessor guts of virtually all personal computers are also divided into two mutually exclusive camps; Apple Corporation uses Motorola ICs, and the rest use Intel or Intel compatible ICs. It should be noted that in common usage the term "PC" now largely signifies IBM and clone machines only, although here we will continue to use it to mean any personal computer.

By 1993 sales of personal computer chips surpassed 30 million units a year. The price has continued to drop while performance has exceeded the wildest dreams of early computer designers. Indeed, a $2,000 Pentium based PC of today can outperform a $30,000 minicomputer of 30 years ago by factors exceeding 100 times, and some are suggesting that PCs are becoming so powerful and inexpensive that the mainframe computer is doomed.

With this computational power at hand at a cost of only a few dollars to a few hundred dollars per chip it's not surprising that a vast number of microprocessor applications have appeared beyond their use in the personal computer. Indeed, not only the microprocessor, but all of electronics has virtually saturated our culture to the point where modern life would come to a dead stop within days without the ubiquitously facilitating silicon chip. Such a sweeping statement deserves more than mere assertion, however, so let's examine a few applications of the transistor and microprocessors to get some explicit sense of our dependence on them.

Manufacturing today would be unthinkable without "CAD/CAM", computer aided design/computer aided manufacturing. Many modern systems essential to our daily activities are so complex that conventional means of design and manufacturing would simply be out of the question. The computer optimized aerodynamics of automobile and aircraft design make possible their efficient operation thereby saving hundreds of millions of gallons of fuel annually. And with the fuel saved there is also a significant reduction in environmental pollution. Microprocessors themselves as well as many other types of ICs absolutely require CAD intervention in their design. It's not difficult to see how this is so when one considers the more than 3 million transistors found in a single Pentium microprocessor chip.

The manufacturing process is often facilitated with the use of electronically controlled robots. Manufacturing done with robots results in a less expensive, more uniform product with tighter tolerances and greater reliability. In a closely related area the machining of complex shapes has been made feasible with electronically controlled machine tools. Some may recall the consternation of the US government when they found that such computer controlled machines had been shipped to the Soviet Union in the mid 1980s thus permitting them to manufacture complex submarine propellers capable of very quiet operation. In an age of nuclear tipped, submarine launched ballistic missiles the detection of submarine noise is an important defense matter, so the foreign acquisition of facilitating electronics to help build low noise subs gives this aspect of electronic technology doomsday overtones. However, similar electronically controlled systems are also used to design and manufacture high efficiency propellers for commercial shipping. The eventual cost savings for shipping is reflected in virtually every imported product we buy.

Process control in manufacturing is an electronic function that is not only a matter of convenience as in the manufacture of beer and gasoline for example, but a matter of survival when one considers pharmaceuticals.

Consumer goods are full of electronics and little computers. Today's typical automobile has electronics in its ignition, antilock breaks, cruise control, air bags, dashboard instrumentation, radio and cassette player, and a host of power accessories. With all this the typical automobile has on board raw data processing power exceeding that of the 1969 Apollo space craft sent to the Moon. The bottom line is that these systems yield substantial improvements in safety and efficiency that save the consumer billions of dollars a year. Similar savings are accrued because of more efficient appliances such as microwave ovens, and automatic setback thermostats. Television sets, camcorders, VCRs, radios, hifis, cordless and cellular phones, and telephone answering machines are obvious products of the electronics revolution, and their impact on our lives in facilitating communication and entertainment is obvious as well. A more subtle and perhaps more profound impact of these devices is reflected in the examples of how rapidly the world got news pictures of the Tiananmen Square massacre in China in 1989 via private FAX transmissions, how the isolationist Iranian government tried and failed to restrict satellite TV reception from abroad in 1994, and how our understanding of law enforcement has been affected by the camcorder record of the Rodney King beating.

Radio and television broadcasting of all kinds from amateur to citizens band to commercial to government broadcasting are one of the essential cores of the information age in which we live. We would have none of this without electronics, and little more without the transistor. A world without radio communications would be a world moving at a 19th-century pace.

The world of commerce can function as it does only because of electronics. Consider for a moment that the stock market sometimes has days where the number of shares traded exceeds half a billion. This would not be even remotely possible without computers doing business and bank accounting, without electronic funds transfer, without computerized market analysis, and without instantaneous international communication. Written correspondence, within or between institutions would also be impossible at today's pace without computerized word processors and their interconnecting networks.

At the grocery store food prices are held down partly by the improved efficiency of automated check out procedures. The laser bar-code scanner coupled to the cash register and the warehouse inventory control system integrate the entire process of getting and selling goods in a timely and profitable manner.

The government's reliance on computers and electronics goes from IRS review of tax records (a painful but necessary burden we must bear for the collective benefits of society) to weather forecasting and climate analysis. Without these aids our agricultural productivity would be diminished, and our personal lives put much more at the mercy of the elements.

The most prominent of all government enterprises is the military establishment. The recent Persian Gulf War has revealed only the slightest hint of what modern military systems can do. Radar, sonar, special radio communications and coordination, aircraft autopilots, precision navigation using inertial navigation systems or the GPS global position system that uses signals from several satellites in space to locate a receiver to within a few feet any place on earth, night vision systems that use amplified low levels of starlight to make the battlefield as visible at night as during the day (and are used as well to catch drug runners crossing the border here at home), smart weapons that can detect, identify, and selectively destroy hidden targets, and cryptography that provides secure communications not only on the battlefield, but also in diplomacy and business.

The space program is another area of government enterprise where electronics is decisively essential and where the rewards are of global importance. Satellite communications are the sinews of international trade and finance. Satellite imagery defines the global patterns of land and sea use and abuse, as well as facilitating the monitoring of the ozone hole at the South Pole. Satellite images are also an important link in discovering global weather patterns and predicting seasonal, continent wide weather trends.

The scientific and educational ventures in space and on earth are simply unthinkable without electronics. No exploration of the planets and the universe could be done, no discovery and understanding of the molecular, atomic, nuclear, and subnuclear properties of matter would be forthcoming, and no engineering of buildings and bridges and dams and power plants. Virtually nothing common in our civilization would exist were it not for the facilitating role of electronics. Computers in the classroom aid students in learning to read, spell, and type. CD-ROM references are becoming common so that immense amounts of data are instantly available on the computer and are often relationally linked to facilitate the students grasp of the coherence of knowledge. So called "multimedia" computer systems are now being to be used to provide unique forms of understanding. Consider, for example, studying Beethoven and his music; a multimedia computer could produce descriptive text on one part of the screen, the score of his Fifth Symphony on another part, and play user selected portions of that score over the computer's audio attachment. Or imagine a student studying Muslim history; the text of the Qu'rān presented side by side in Arabic and English with selectively highlighted Arabic words explained in a third portion of the screen while the audio gives the pronunciation.

In medicine the world of computers and electronics surrounds us in the operating room and later in the intensive care room. Monitors watch our vital signs while complicated electronically controlled machines substitute for our lungs and heart. During the diagnostic phase we are examined with ultrasound, MRI imaging, PET and CAT scans, and old reliable X-rays. Much of the imaging is enhanced and made sharper and clearer with extremely sophisticated computerized image processing. If necessary, electronic pacemakers can be used to help heart patients, and of course hearing aids are now so small they can fit completely in the ear or in the temple piece of ordinary glasses. There are even medical electronic devices that can be taken home so that we can monitor our blood pressure, and our blood sugar level (without that troublesome need to draw blood). Meanwhile, "expert programs" on the physicians computer are aiding the diagnosis of rare and obscure conditions that might otherwise be a mystery.

In the home, smoke detectors, carbon monoxide detectors, and radon detectors are contributing to the reduction of the normal hazards of life. Electronic security systems reduce another hazard around the home and are also found almost everywhere in the business sector. We are living twice as long as our ancestors, due in no small part to the electronic advances of the 20th century.

This smallest of samples should persuade us that the role of electronics and its impact on our times cannot be overestimated. The invention of the quintessential electronic component, the transistor, marks an achievement comparable to the most influential in human history ranking along side the first use of fire, the invention of the wheel, and the invention of movable type. We have seen that electronics has given us individually and collectively superhuman capabilities. Benjamin Franklin, the father of it all, would think us gods were he to visit us in our century.

Chapter 15

Nuclear Science & Technology

If "revolution" means rapid and fundamental change then the 10 years from 1895 to 1905 are as revolutionary as any ever seen in the history of science. Virtually every modern concept regarding the structure of matter had its origins in that decade. Wilhelm Roentgen in Germany discovered X-rays in 1895 (Nobel-Physics, 1901), Henri Becquerel in France discovered radioactivity in 1896 (Nobel-Physics, 1903), J.J. Thomson in England discovered the electron in 1897 (Nobel-Physics, 1906), Max Planck in Germany discovered the quantum nature of radiation in 1900 (Nobel-Physics, 1918), and Albert Einstein in Switzerland discovered Special Relativity and explained the photoelectric effect in 1905 (Nobel-Physics, 1921.)

Roentgen had been following a line of research pursued rather fruitlessly by others for more than 20 years, but as genius so often does, he discovered the essence of the matter. Like many others, he observed the luminescence appearing on the glass envelope of his cathode ray tube (the same basic device as a TV picture tube) as high voltage was applied. He was also able to show that the glow was produced by penetrating rays that could even expose photographic film wrapped in black paper and at some distance from the apparatus. In addition, he found that these rays could penetrate his hand and cause a shadow of his bones to appear on a screen of phosphorescent material. Since these rays were heretofore unknown, he called them "X-rays."

Becquerel heard of Roentgen's work and was curious to know if X-rays might not be produced by the action of sunlight on certain crystals. By chance, one of the materials he was trying was potassium uranyl sulfate, and by further chance, he happened to store some of it for several days with an unexposed photographic plate wrapped and protected from light. The potassium uranyl sulfate and photographic plate had been put together as part of the test with X-rays and bright sunlight. Overcast days had not allowed him to complete the experiment, so he went ahead and developed all the film at hand. To his astonishment he found the images of the chemical crystals brightly burned into the developed film. Others, as far back as 1867 had observed similar results, but it was Becquerel who discovered that sunlight had nothing to do with it, and that the exposing agent was emanations from the uranium compound itself. Thus was radioactivity discovered (though the term, "radioactivity," was coined two years later by Marie Curie.)

Meanwhile, Thomson like Roentgen, had been experimenting with a cathode ray tube, but he was more interested in the cathode rays themselves rather than the newly discovered X-rays. German scientists believed that cathode rays had something to do with the conjectured basic stuff of all space, the aether. The English scientist William Crooks had earlier even speculated that cathode rays represented a fourth state of matter beyond gas, liquid, and solid, the "ultragaseous" state. Thomson, however, living up to his reputation as a very careful experimenter, avoided the errors of his predecessors and determined that cathode rays were negatively charged particles very very much lighter than hydrogen and that they were a basic constituent of all matter. The name given to the particle, proposed a few years before, was "electron."

In a completely different area of physics Max Planck was trying to understand the process of absorption and emission of heat radiation from an ideal material called a black body. The best minds in 19th-century physics had developed only partially successful theories; one worked at low temperatures but failed at high temperatures, the other worked at high temperatures but failed at low temperatures. In all cases the assumption was that heat radiation absorption and emission could occur in any amount whatsoever, that is, that it was a continuous phenomenon. This assumption sounds reasonable enough, even today, but it is profoundly wrong. In one of the most daring intellectual achievements of all time Planck assumed that heat radiation could be absorbed and emitted only in discrete energy packets now called quanta. The energies of the individual quanta are just equal to a constant ("Planck's Constant") times the frequency of the radiation (in the visible domain, the color). With this assumption he very successfully established the quantum nature of radiation and planted the first seed of what would become the entire field of quantum physics, the only physics that has successfully given us our understanding of the molecular, atomic, and subatomic nature of matter.

Einstein got his undergraduate degree only a few months before Planck announced his discovery of the quantum of radiation. Five years later, in 1905, Einstein was to publish three papers, any one of which was easily worth a Nobel Prize. His eventual Nobel was for a theoretical interpretation of the photoelectric effect. The photoelectric effect is observed when electrons are knocked out of certain metals as light shines on them. What was perplexing before Planck and Einstein was that the threshold for electron emission depended not on the brightness, but on the color of the light with the violet end of the spectrum being more efficacious than the red end. For example, if a particular metal had a photoelectric threshold for green light, then no light from green through the spectrum to yellow to orange to red would cause electron emission; and this is so, no matter how bright the light. So for that particular metal one could shine the brightest conceivable yellow light on it and yet get absolutely no electric current from it. On the other hand, given the green photoelectric threshold, one could shine the dimmest green or blue or violet light on the metal and always get a current. It was this color dependence of the threshold and the absence of a brightness threshold dependence that mystified everybody. Einstein took Planck's idea of a quantum of heat radiation in emission and absorption processes and extended it to transmissions through space. He also understood that light radiation was the same as heat radiation, differing only in its higher frequency. (Radio and TV waves are the same thing as light and heat waves only they have a lower frequency.) Now the photoelectric effect was easy to understand; the individual quanta of higher frequency green, blue, or violet light have more energy, indeed, they have enough energy to overcome the energy barrier at the surface of the metal and cause electrons to escape. Quanta of lower frequency yellow light cannot impart enough energy to the electrons in the metal to overcome the surface barrier, and so no current flows, and this is so, no matter how many yellow quanta hit the metal, that is, no matter how bright the yellow light. This theory of Einstein's not only explained the photoelectric effect but also reestablished the particle nature of light that had fallen into disfavor during the 19th century with the very successful wave theory, and is one of the seminal sources leading to the eventual understanding of the dualistic (particle/wave) nature of radiation and matter.

The "miracle year" of 1905 ending the revolutionary decade,1895 to 1905, has been so labeled because of Einstein's scientific production. His 1905 work has had a prodigious and profound impact on science for the whole of the 20th century. For our story of the impact of nuclear science and technology on the century the key discovery was the equivalence of mass and energy developed out of his Special Theory of Relativity and embodied in the celebrated equation, $E=mc^2$. Since I have already touched on Special Relativity in Chapter 13 we shall pass on, recalling only that nuclear weapons, nuclear power, and nuclear medicine all owe their existence to this fundamental understanding.

The pace of discovery in the next decade can be characterized more as a time of application of theoretical ideas of the preceding decade than a time of deep new insights. Two players stand out here for their development of the atomic model, Ernest Rutherford, and Niels Bohr.

Rutherford already had a Nobel Prize in chemistry for his work on radioactivity. (This amused him to no end since he was a physicist, not a chemist, and virtually everyone but the Nobel Committee acknowledged the work as "physics.") The best model of the atom at the time was a "plum pudding" model envisioning a sphere of some sort of electrically positive stuff with the negative electrons Thomson had discovered embedded in it. Beginning in 1906 Rutherford began finding evidence that this could not be the case. His experiments showed that the atom, far from being a sphere of something positive with negative electrons uniformly distributed throughout it, had a very small, very massive central nucleus with the electrons located a relatively great distance away. He later guessed that the nucleus must have a positive charge. His findings, published in 1911 and 1912 gave us our first modern picture of the atom. Nevertheless, as good as it was, it had a serious problem. Rutherford's nuclear atom couldn't possibly be stable. It would destroy itself in a small fraction of a microsecond. Since atoms obviously didn't do this (witness that we are all here and full of atoms) the model needed work.

Neils Bohr, whose influence on 20th-century physics is comparable to that of Einstein's, obtained his PhD in May, 1911 in Denmark. A few months later he left for England to work first with Thomson then with Rutherford, and returned to Denmark within a year, his head full of ideas about the nature of the structure of the atom. In July 1913 with only his fourth publication he defined the atom largely as we know it today. Bohr took the ideas of Planck and Einstein, and the experimental evidence of physicists studying the emitted spectra of hot atoms (we see similar spectra of hot atoms in the sun when sunlight passes through water droplets in the air creating a rainbow) and proposed the quantum atomic model. This model not only harmonized with the latest theoretical work but solved the problem of Rutherford's unstable atom and accounted for the abundant spectroscopic data at hand. Extending the newly established quantum notion of radiation to the atomic model, he envisioned an atom where the electrons moving around the nucleus must occupy only certain "allowed" orbits each with its specific energy level, and that any electron transition between orbits required either the emission or absorption of energy equal to the difference in the orbital energy levels. Electrons simply were not allowed by nature to spiral into the nucleus and destroy the atom as would be the case in the classical Rutherford atom. The electrons were stuck in their particular orbits until something gave them the necessary energy to go to another orbit, as for example, when an atom is heated. The importance of Bohr's insight cannot be overestimated. It is the foundation upon which all subsequent understanding of atomic physics is based and was immediately recognized by physicists around the world as a major discovery. Neils Bohr received the Nobel Prize for physics in 1922.

In the 1920s and 1930s a large number of important and profound discoveries were made in the theory of the fundamental structure of matter, as alluded to in Chapter 13, but our interest here is limited to specific developments with the atom and its nucleus. Accordingly, the next relevant event occurred in 1932 when the Englishman, James Chadwick discovered the neutron.

As far back as the early 1920s suspicion was growing on the part of Rutherford and others that a heavy neutral particle might exist in the atomic nucleus along with the already known positive proton. Rutherford and his associate James Chadwick searched for it but found nothing. However, some years later, radiation yielded by the experiments of Irene Joliot Curie and her husband Frederic Joliot and others, which they misinterpreted as gamma rays, was all that Chadwick needed to clinch the case for the neutron.

The neutron turned out to be a wonderful tool for studying atomic structure. With a mass slightly greater than a proton it is heavy enough to smash into atomic nuclei and create all kinds of interesting results. Since it is electrically neutral it can approach the nucleus without the interfering effects of the electric charges of the protons or the surrounding electrons. Moreover, a source of neutrons is easy to produce; one typical method is to add some beryllium metal to radium. The radium radiation interacts with the beryllium and produces neutrons. Among the first to use these properties was an unusually gifted young physicist in Italy named Enrico Fermi.

Enrico Fermi was, arguably, the most broadly gifted scientist of the 20th century. It is very unusual to find a scientist with both top rank theoretical and experimental skills. Stories abound of Nobel theorists whose experimental abilities were not only nonexistent but actually seemed to be malevolent, and of Nobel experimentalists who were virtually helpless in the face of theoretical physics. But Fermi was a different sort of man, on the one hand he was noted for the lucidity and simplicity of his theoretical insights and on the other, he marshalled the same talent for simplicity and directness in service of his considerable experimental gifts. Many are familiar with the story of his on the spot estimation of the yield of the first atomic explosion at Alamogordo, New Mexico in 1945 when he simply let strips of paper fall to the ground during the passing blast wave. Observing how far the strips were carried he could estimate the properties of the blast wave and thus the energy of the source, the atomic explosion. His answer was 10 kilotons of TNT equivalent; more precise measurements and calculations later gave 18.6 kilotons. (Preblast guesses on the probable yield ranged from 0.3 to 45 kilotons.) However, Fermi's talents went further, for he was a gifted teacher as well. At the University of Chicago where he worked before his premature death from cancer in 1954 he was famous for his freshman physics lectures. Indeed, students in the humanities who otherwise had no interest in science would fight to get a place in his class, because he, "...made physics so understandable."

After Chadwick's discovery of the neutron Fermi quickly realized its potential for producing nuclear reactions. In Rome he began a series of experiments irradiating more than 60 elements with neutrons of various energies. In the process he created many new radioactive elements, and discovered, by merest chance, that slow neutrons, could on occasion, be hundreds of times more effective in creating radioactive elements than the high energy neutrons he had been using. In less than four years these discoveries were recognized by award of the 1938 Nobel Prize in physics.

Fermi's experimental program was very systematic. His plan was to start with the lightest element and work completely through the list of known elements till he reached the heaviest, uranium. By the time he reached uranium in 1934 he had what appeared to be a fairly good working hypothesis about what was happening as each element responded to neutron radiation. The target nucleus would absorb a neutron, become unstable and emit radiation, and finally stabilize as a new element slightly heaver than the original target. In some cases it was also possible for the transformation to go in the other direction towards a slightly lighter element. When he irradiated uranium the results were ambiguous; did he have a heavier or a lighter final product? The physics suggested a heavier, transuranic element, and since the chemistry eliminated the possibility of slightly lighter elements, Fermi supposed that he had discovered the first man-made elements heaver than uranium, elements 93 and 94. He even named them; Ausenium and Hesperium. He was, however, wrong; he produced nothing of the kind. What he had done, without knowing it, was split uranium into two nearly equal parts. He had achieved uranium fission, one of the fundamental processes for all nuclear power generation and nuclear weapons. However, the conscious discovery of uranium fission would have to wait another four years, but when it came, everybody knew what it meant, and the portent was not good.

Meanwhile at the Kaiser Wilhelm Institute in Berlin Otto Hahn, Lise Meitner, and Fritz Strassmann had also been intrigued by the problem of slow neutron induced uranium activity. They too observed results that they couldn't explain. By 1938 they were confident that no transuranic elements were being produced, indeed, the chemistry pointed to elements lighter than uranium. Hahn, whose experimental abilities were unsurpassed, was finally forced to conclude that the daughter element produced by uranium absorbing a slow neutron was barium, but this was impossible; how could the absorption of a slow neutron cause a uranium nucleus to split virtually in half? It was with extreme reluctance and great caution that Hahn and Strassmann (Meitner had since fled Hitler's terror) announced their results in January, 1939. Prior to publication Hahn exchanged a number of letters with Meitner, now in Stockholm, pleading with her to come up with some theoretical explanation that would avoid having to conclude that the product was barium. It was not to be; this was a case of 'it quacked like a duck; it walked like a duck; and it looked like a duck.' This duck was indeed barium.

Lise Meitner and her nephew, Otto Frisch, struggled with the barium result just as Hahn was doing, but finally they accepted the results and addressed the theoretical consequences. If the product of uranium absorption of a neutron was barium, then the uranium atom must have fissioned, i.e. split nearly in half. This was an entirely new nuclear process, and there was more; their simple calculation also showed that an immense amount of energy must be released in the process, tens of millions of times that which would be released in a comparable chemical reaction.

Within a matter of days the news of uranium fission spread across the European and American physics communities. Everyone instantly understood the potential consequences of this discovery, the atomic bomb. It was mid January, 1939.

Enrico Fermi had received his Nobel Prize and very judiciously decided to immigrate to the United States and not return home to fascist Italy. When he heard the news at Columbia of uranium fission he understood with some embarrassment that his 1934 work was partially flawed. We can see an artifact of this in his Nobel Lecture where he catches the text just before going to the printers and adds a footnote acknowledging the discovery of fission. Thereafter events move rapidly; Fermi and colleagues, essentially duplicating experiments done just days before in Europe, confirm not only uranium fission but its gigantic energy yield. By October, 1939 World War II had begun in Europe, and President Roosevelt had received Einstein's famous letter warning of the potential of atomic weapons.

Under the pressures of war the next six years saw the focus of nuclear science temporarily change from basic physics to a sort of applied physics. Physicists and their governments in England, the United States, Germany, Japan, and the Soviet Union all knew of the possibility of making atomic bombs and all launched programs of varying ambition. However, in a preview of what was to come in the following decades only the United States and the Soviet Union developed a substantial national effort to develop the bomb. Roosevelt's positive response to Einstein's letter began with a modest $6,000 appropriation and the formation of a committee to explore the matter. It culminated with the $2 billion "Manhattan Project."

The Manhattan Project was the code name for the atomic bomb project. It had the highest priority of any program in the war effort, a necessary condition since it consumed a significant amount of the nation's resources. On the material side the plant to produce the weapons form of uranium, U-235, at Oak Ridge, Tennessee was as massive as any found anywhere. It literally created a city for its support. Elsewhere, at Hanford, Washington, another almost as impressive a facility was built to produce plutonium, the other bomb grade material. The intellectual resources recruited were equally as substantial. In the beginning Enrico Fermi gathered a group of scientists at the "Metallurgical Laboratory" at the University of Chicago to determine whether a self sustaining nuclear chain reaction could in fact be achieved. Not long thereafter the principal weapons design and construction laboratory was founded at Los Alamos, New Mexico.

By 1940 there was no doubt that a fundamental process existed to release atomic energy, but all that was known was that it could be released one atom at a time. This was a stupendous amount of energy for only one atom to release, but on a human scale it was hardly noticeable. For example, the energy release from the fissioning of one uranium atom is about 200 million electron volts (chemical reactions are about five electron volts); this is enough to move a grain of sand a barely perceptible amount. So the question was, how to make a lot of uranium atoms fission at virtually the same instant. Fortunately, Leo Szilard, a seemingly gypsy physicist who never settled down anywhere, had envisioned the possibility of a chain reaction, that is, where the first generation of reacting atoms would initiate a second generation of reactions which in turn would start a third generation, etc.. Since it was neutrons that started uranium fissioning, and since it had been discovered that neutrons were among the reaction products along with the barium that had so perplexed everyone at first, then it just might be possible to use these secondary neutrons to sustain the reaction in a chain of successive generations of fissions; a self sustaining nuclear chain reaction.

That is the concept, but there were some not so minor questions that had to be answered before the idea of a bomb could become anything more than a wild idea. For example, how many secondary neutrons were produced per uranium fission, and how many of those produced were lost without interacting or otherwise failed to contribute to the chain reaction. So the Metallurgical Laboratory was assembled at the University of Chicago to find the answers. Fermi and coworkers, most notably the young Herbert Anderson who was the first on this side of the Atlantic to observe fission, constructed the CP-1 (Chicago Pile Number One). As the first nuclear reactor the CP-1 was the progenitor of all research and power reactors to follow. CP-1 was a pile of graphite blocks with uranium imbedded in it. The graphite slowed the neutrons to facilitate the fission process. Constructed in a squash court under the west stands of the old Stagg Field on University Avenue, it made history late on the very cold, bleak afternoon of December 2, 1942 when the carefully controlled chain reaction became self sustaining. The Atomic Age had arrived. Also, The Bomb was now a virtual certainty.

Today Stagg Field has been relocated a few blocks away, and the stands have long since been removed (inspired in part by college students snatching souvenir bricks which were mildly radioactive). The site is now part of the west lawn of the Regenstein Library, and the event is commemorated by an eerily symbolic bronze sculpture by Henry Moore entitled "Nuclear Energy." Although the sculpture is at once very evocative of a mushroom cloud or a human skull, Moore actually independently designed the abstract piece well before being approached by the 25th Anniversary Commemorative Committee in the mid 60s.

Los Alamos today is an easy 40 minute drive northwest of Santa Fe via a fine divided highway. Not so in 1942, the only destination after an arduous two hour journey up into the hills of the Jemez Mountains was a private school for boys, the Los Alamos Ranch School. J. Robert Oppenheimer, the newly appointed director of the planned weapons laboratory, selected the boys school site only a couple of weeks before the historic CP-1 experiment. Remoteness and accessibility were among the important selection criteria. Remoteness, to more easily control security in a community that would have several hundred very verbal scientists, and accessibility, to facilitate supply of a laboratory that would require considerable hardware resources from around the country including several "atom smashers." Site Y, as Los Alamos Laboratory was originally known, was fully operational within five months. The laboratory's mission was to design and build an atomic bomb. In order to do this considerable theoretical and experimental work had to be done to determine how much uranium or plutonium would be needed, as well as determine how to initiate the explosion. It turns out that 33 pounds of uranium is required for a critical mass, about the size of a grapefruit. The same job can be done with a tennis ball size, 11 pounds of plutonium.

It is interesting to note that although the hydrogen bomb would not be built and successfully tested for another 10 years, the theoretical possibility was recognized early on in the Manhattan Project and was one of the secondary goals of the program. Edward Teller, later known as the "father of the hydrogen bomb," was to spend most of his time at Los Alamos thinking about the "Super" (as the hydrogen or fusion bomb was then called.) Nevertheless, the Super was clearly a distant second to the fission bomb in priority and received little attention until well after the war.

The hydrogen bomb differs from the uranium or plutonium bomb in the nuclear process of releasing atomic energy. Uranium and plutonium nuclei at the heavy end of the atomic scale split roughly in half, they fission. The mass of the parts is less than the mass of original whole. The mass difference is manifest as energy in accordance with Einstein's $E=mc^2$, which, with the special circumstances of bomb design, results in an explosion. At the other end, the light end of the atomic scale we see the same principal in action. This time there is no splitting, there is a joining, a fusion, of hydrogen nuclei. The mass of the new nucleus, helium, is less than the mass of the two hydrogens; the missing mass equals - boom! Nuclei at both ends of the atomic scale show this mass excess, which in some cases can be released as energy, but as one goes gradually to the center of the scale, towards elements like iron, the excess disappears and we have the most stable elements.

Most scientists expected their atomic bomb to have a yield equivalent of a few thousand to a few tens of thousands of tons of TNT. However, by 1944 it was already clear that improvements in the bomb's design could eventually increase the yield to as much as 1 million tons TNT equivalent. Furthermore, Teller was proposing that a Super might be built with a yield of 100 million tons of TNT.

Clearly the almost hysterical plunge to increase the destructiveness of atomic weapons existed well before the arms race with the Soviets. The extent of this hysteria was, and is, partially hidden by the facility we've all developed in discussing the magnitude of atomic explosions, bombs with yields of 10,000 tons (10 kilotons),
1 million tons (1 megaton), 100 million tons (100 megatons). Easy to say, but just what are we saying? Perhaps it might serve us better if we defined these bombs in terms of the size of the TNT pile giving the equivalent explosive yield. Ten kilotons of TNT would form a pile 57-feet high. That's about equal to the height of a six story building. One megaton would make a pile of TNT 266-feet high, and 100 megatons a pile as high as the Empire State Building, or more than twice the height of the Great Pyramid of Giza. The total amount of high explosives used in World War II was about three megatons. The biggest hydrogen bomb exploded by the Soviets was said to be about 50 megatons, or 17 times the WW II total. One bomb.

The first atomic detonation occurred just before dawn on July

16, 1945. Oppenheimer had chosen the name "Trinity" for the test. The site was a plane called Jornada del Muerto - the journey of death - 200 miles south of Los Alamos and 60 miles northwest of Alamogordo between the Rio Grande River and the Sierra Oscura Mountains.

Heavy rains the night before put the test date in doubt, but eventually the downpour and thunder and lightning stopped, and less than 1,000 days after Fermi's CP-1 experiment a new thunder and light burst upon the earth characterized by one author as "brighter than a thousand suns." The line is from the Hindu poem, the *Bhagavad Gītā* (The Celestial Song) which Oppenheimer knew well in the original Sanskrit. He said later that the nuclear fireball evoked the Gītā words of the god Krishna asserting his power, "I am become Death, the shatterer of worlds." The scientists who witnessed the Trinity detonation returned to the Laboratory visibly awed, as if they had indeed been instructed by Krishna.

The world shattered for Japan on August 6th at Hiroshima and August 9th at Nagasaki. Two bombs, one of uranium, 15-kiloton yield, the other of plutonium, 22-kiloton yield. A third of a million dead from prompt and delayed effects, two cities destroyed, one war promptly ended.

The end of World War II did not, however, end the development of atomic weapons. Teller continued to push for development of the Super, but demobilization was in the air, and he got little sympathy at first. Nevertheless, his argument about the growing Soviet threat soon gained currency and the Cold War replaced World War II. The examples of Hiroshima and Nagasaki were, arguably, sufficient to discourage serious confrontations of the nuclear powers which otherwise might have succumbed to the traditional historical machismo that has so often led states into war. In any case there was no overt war, just the feverish preparations for one, at least at the nation's nuclear laboratories.

The Soviet Union exploded its first atomic bomb in September, 1949; the United States exploded its first hydrogen "device" in October, 1952. With a yield of 10.5 megatons it not only vaporized the little island of Elugelab on which it rested but left a hole in the Enewetak atoll ocean floor 164-feet deep and well over a mile in diameter. The Soviets followed suit with their own hydrogen "device" in August, 1953. (They were called "devices" because they weren't yet weaponized into a deliverable bomb.)

With the nuclear arms race in full stride Los Alamos once again had a mission, which it now shared with the newly created weapons lab, the Lawrence Livermore Laboratory at the University of California. (By the way, Livermore should not to be confused with the University of California's Lawrence Berkeley Laboratory. The Berkeley lab is devoted entirely to the study of fundamental physics and rather ardently disowns its cousin at Livermore.) The cachet of scientists, especially physicists, was firmly established by the success of the Manhattan Project. Government support of physics research was beyond the wildest dreams of its practitioners. It continued so until the Vietnam War in the late 1960s became culturally so costly that we found we definitely could not have "guns and butter" at the same time. The "butter" of physics took a hit from which it is still suffering.

However, this was not the case with weapons physics. The Cold War continued to drive nuclear weapons development at a breathtaking pace through the 1950s and 1960s. The momentum established by the nuclear weapons program was so great that even today, almost a decade after the evaporation of the Soviet threat, development and testing continue.

At the Los Alamos National Laboratory a new culture was evolving, cloaking itself in the flag of "national defense" and assuming all the trappings of a secret society. Any program was acceptable if national defense warranted it. Fiscal and scientific oversight became progressively less disciplined, frequently hiding behind a wall of secrecy. It became possible for mediocre technical work to be shielded from adequate peer review by classifying reports with a security rating so high that few could have access to them. After awhile this intellectual distortion twisted to the point that within the defense community the quality and importance of a report was primarily determined by the level of its security classification. A "TOP SECRET," "RESTRICTED ACCESS," or "SENSITIVE NUCLEAR MATERIAL" classification was automatically the imprimatur of quality and importance, and reports with lesser classifications, or none at all, were automatically assumed to be inferior. It's not surprising that as this professional attitude developed it permeated the social realm as well, where at Los Alamos one's social standing was largely determined by the level of one's security clearance, and therefore one's importance.

The independent little world of nuclear weapons work felt it had license to do what it pleased and an accommodating American people declined to ask many penetrating questions, if they asked any questions at all. Recent revelations show where such neglect has led. The disposal of radioactive materials at nuclear weapons production facilities, especially in Colorado and Washington State, has been cavalier at best with little apparent regard for the welfare of the populations in the areas. Even more arrogant has been the test exposure of individuals and populations to plutonium. Government nuclear scientists secretly injected plutonium into individuals and released it into the environment to study its dispersal and effects on populations.

Weapons research led not only to the atomic and hydrogen bombs, but to active consideration and sometimes development of other types of bombs as well. In 1950 some feared that even a cobalt bomb might be produced. It was one of a class of weapons studied made deliberately "dirty", ie., lots of radioactive fallout. It was to be a conventional nuclear weapon containing a great deal of the common metal, cobalt-59. Neutrons from a nuclear explosion would change the innocuous cobalt-59 into the highly radioactive isotope, cobalt-60. The cloud of cobalt-60 would then sweep over the enemy irradiating them with its high energy gamma rays, but there was a problem with this idea; cobalt-60 has a half life of five years. That means that of any given quantity of cobalt, half will decay in five years, and half of the remainder will decay in the next five years, and half of that remainder will decay in the following five years, etc. So after the cobalt cloud passes over the enemy it will continue around the world passing over us and many of our friends as well, continuing for decades. Clearly, the lack of control was, as it were, fatal to the idea.

Sam Cohen was a physicist working for the Rand Corporation when he came up with the idea of an "enhanced radiation weapon" in 1958. We know it as the neutron bomb. A neutron bomb is basically a very small hydrogen bomb where the fusion reaction is specifically tailored to yield a large fraction of its energy in high energy neutrons. A one kiloton neutron weapon has a "military radius" of three-quarters of a mile. This means that the neutron radiation at that distance would almost instantly incapacitate a person and very shortly later kill him. By way of comparison, lethal radiation doses usually take from days to weeks to months to kill by the slow disintegration of the body's cells and functions.

The subtleties and fine points of nuclear weapons design have been carried to a high art; weapons have been designed to produce unusual amounts of X-rays, to maximize blast or heat, to produce stupendous electrical pulses that disable the enemy's electronics (the so called EMP weapons), and to be very small. The latter have been fitted into six-inch artillery shells, and into suitcase-like systems for nuclear demolition in noncombat situations.

The development and testing of nuclear weapons continues although it has declined sharply in the mid 1990s. Today Los Alamos National Laboratory and the Lawrence Livermore National Laboratory are beginning to look for other, non-nuclear missions. Nevertheless, the 'Dr. Strangelove' legacy of a world full of nuclear weapons variously estimated at 20 to 30,000, many distressingly portable, will be with us long into the 21st century. However, nuclear science and technology are not by any means limited to weapons. There are numerous beneficial applications, two of which are radiology and nuclear power.

Medicine began using the discoveries of nuclear physics even before there was knowledge of the nucleus. Recall Wilhem Roentgen's discovery of X-rays in 1895 and his observation of the photographic image of the bones in his hand. The medical value of this ability of X-rays was recognized and used within months. The first medical use of X-rays in the military occurred in 1898 after the battle of Omdurman in the Sudan. (The casualties were taken in the last calvary charge of the British Army, in which the young Winston Churchill was an unwounded participant.)

From the very beginning radiology found applications in two broad areas, diagnosis and therapy. We are all familiar with the use of X-rays in diagnostic radiology and in dentistry; more than half of the population gets some kind of X-ray every year. The use of X-rays changed very little during the first half of the 20th century awaiting the innovations in electronics and nuclear technology spurred by World War II. The situation was similar in therapeutic radiology where X-rays were found to be effective in destroying cancer cells faster than healthy cells. In the meantime, radium had been discovered by Marie Curie and her husband Pierre in 1898, for which they received the 1911 Nobel prize in Chemistry. The first radium therapy was undertaken in 1904 where radium needles were placed under the skin next to tumors, but it wasn't until 1909 that it was discovered that tumors were in fact more sensitive to radiation than healthy tissue. Radium therapy continued pretty much unaltered until more effective and safer radioisotopes came into use after World War II.

After 1950 radiology came into its own; not only did it have its diagnostic and therapeutic areas, but it eventually further subdivided into radiology and nuclear medicine. Nuclear medicine largely concerns itself with procedures that require the use of radioactive elements, whereas radiology is generally involved with the use of X-rays; however, for simplicity I will use only the term, "radiology."

Advances in electronics, and later the appearance of low cost computers, were particularly important in the development of an entire family of imaging devices. Scanners and gamma ray cameras, invented in the 1950s, were common in hospitals by the 1970s. These devices detect gamma rays coming from radioisotopes which are administered to patients in a chemical form that favors their concentration in the organ of interest. For example, gamma emitting iodine-131 administered to a patient concentrates in the thyroid. A gamma ray scanner or camera detects the gammas and presents a kind of picture of the thyroid, indicating by the brightness of the image whether it is normal, or over or under active. Similar techniques are used to image the heart, liver, brain, and many other organs. The diagnostic radiologist's job is to select the proper radioisotope and decide how much to use based on the organ to be imaged, its biochemistry, and the half life of the isotope. (Isotopes are different forms of an element, chemically identical, but physically heavier or lighter; when an isotope is radioactive it's called a radioisotope. See Table 1.)

Since the 1970s far more advanced systems working on slightly to substantially different principals have dominated the diagnostic imaging field. These include the CAT scanner, the PET scanner, and the MRI scanner. All three give very impressive pictures of internal organs and their pathologies, but the MRI scanner has the

Radioisotope	Source	Half Life	Significance
tritium (hydrogen-3)	man made	12 years	weapons, glowing signs, fusion power?
nitrogen-13	man made	10 minutes	medicine (PET scans)
carbon-14	natural	5730 years	archeological dating
potassium-40	natural	1.26 billion years	geological dating
cobalt-60	man made	5.26 years	medicine (radiation therapy)
strontium-90	man made	27.7 years	dangerous fission byproduct
technetium-99m	man made	6 hours	medicine (ideal for diagnostic radiology)
iodine-131	man made	8 days	medicine (therapeutic radiology) fission byproduct
cesium-137	man made	30 years	dangerous fission byproduct
uranium-235	natural	700 million years	weapons and nuclear power
plutonium-239	man made	24,000 years	weapons and nuclear power

Table 1

Some Important Radioisotopes

advantage that it is "noninvasive," that is, no harmful radiation is involved. (Even the minimally harmful doses from the other procedures are best avoided where possible.) Modern medical imaging instrumentation has come a long way since the days of primitive X-ray tubes creating pictures not much better than crude shadow graphs of bone and soft tissue on film. Part of the advance has come from improved sources of radiation, and a major part from the way the radiation is detected and processed.

CAT (Computerized Axial Tomography) scanners use conventional X-rays, but instead of detecting a single image on film they electronically detect and computer process multiple images from a large number of positions around ("axial") the patient. This allows the computer to reconstruct a slice ("tomography") of interior structure in great detail.

PET (Positron Emission Tomography) scanners operate very much like CAT scanners except for the source and type of radiation. In this case the radiation originates from radioisotopes ("positron emitting") injected into the patient. The emitted positron typically travels a fraction of a millimeter and then encounters an atomic electron in the patient's tissue. When it does, the positron (anti matter) and the electron (matter) annihilate each other creating a pair of gamma rays that go off in opposite directions and are detected by the PET scanner.

MRI (Magnetic Resonance Imaging) depends on the fact that atoms respond in identifiable ways to various combinations of radio waves and magnetic fields. A magnetic field is arranged that is different in each part of the area to be scanned, for example in a brain scan the magnetic field might be stronger at the front of the head and weaker at the rear. Pulsed radio frequencies are created and their interaction with the magnetic field and the tissue of interest creates a map of the tissue density. Sometimes even the health of the tissue can be established and mapped. The detected signals are computer processed in the same fashion as CAT and PET scans, but the resultant image is very much better.

One problem with these new systems is that they are all very expensive, typically well over $1,000,000, so the cost to the patient is considerable, and the temptation on the part of the hospital to prescribe their use in order to pay for them is very strong.

In therapeutic radiology the development of nuclear reactors evolving directly from the original CP-1 facilitated the production of a wide variety of useful radioisotopes, most prominent of which is cobalt-60. Cobalt-60 contained in a large lead vessel with a small hole in it lets out the high energy gamma rays in a well defined beam that can be precisely directed at a tumor with a minimum of damage to nearby healthy tissue.

Another advance from war work was the development of the linear accelerator, a kind of atom smasher that speeds up a beam of electrons to energies of many millions of electron volts. These can then be directed at cancer tissue just as one does with the cobalt machine.

A more controversial application of the discoveries in nuclear science is nuclear power. In a sense, nuclear power predated nuclear weapons, for CP-1 was operationally approaching its designed power output of 200 watts of heat two and a half years before the detonation of the first atomic bomb. Of course that was incidental heat and not what CP-1 was designed for. The potential for the useful use of nuclear energy was recognized at least as far back as 1933 when Leo Szilard conceived of the effectiveness of a neutron chain reaction. However, at the time he had no idea what element might facilitate this. Uranium fission, of course, was the answer, but that wouldn't be recognized for another five years, and when it was recognized it was quickly swallowed up in weapons programs.

Immediately after the war there was some civil interest in nuclear power, but the radioactive fallout from a vigorous above ground weapons testing program began to alarm citizens. This in turn began to place in perspective the possible consequences of a nuclear reactor accident. The power utilities and insurance companies were so concerned by these matters that they wanted no part of nuclear power. On the other hand, the government was anxious to see this technology developed, and did what it could with legislation and demonstration programs to facilitate it. This and the end of above ground nuclear weapons tests eased matters so that the public and the power utilities industry became less concerned and development of nuclear power began in earnest.

Although Great Britain and the Soviet Union seriously

committed to nuclear power before the United States, much of the important development and utilization occurred in the U.S. Perhaps the beginning can be traced to the experimental breeder reactor, EBR-1, at the National Reactor Testing Station in Arco, Idaho which in 1951 had some of its excess heat used to produce a small amount of electric power. About two years later a prototype submarine reactor was tested at the same place leading to the launch in 1954 of the Nautilus, the first nuclear powered submarine. Shortly thereafter the first Soviet nuclear electric power plant began operation with a 5-megawatt capacity. Like the EBR-1 it was a dual purpose reactor, in this case its primary function was to produce plutonium. Elsewhere, the Soviets launched the first nuclear powered surface vessel in 1957, the icebreaker, Lenin.

In 1957 the U.S. government joined with private industry to produce the first nuclear power plant dedicated to producing only electricity. The Shippingport, Pennsylvania facility had the same basic design as the Nautilus reactor, albeit much larger, and came on line with a 60-megawatt electrical capacity. The Shippingport design proved to be such a sound design that it produced power for 25 years, and became one of the prototypes for the entire U.S. nuclear power program.

Nuclear power for the civil maritime industry, however, was not as successful. In 1962 the Savannah, the first nuclear powered merchant ship was launched, but it never became commercially viable because of its expensive reactor technology and the scarcity of qualified operating personnel. The Savannah was scrapped after only a few years of operation.

By the 1970s commercial nuclear power was becoming economically competitive, and it appeared that the era of electric power, "...too cheap to meter" might be coming into view. The number of commercial nuclear power plants in the United States began to rapidly increase with fifteen in 1970, seventy four in 1980, and one hundred twelve in 1990. Similar growth occurred abroad, especially in Great Britain, France, the Soviet Union, and Japan.

However, all was not well with the nuclear power industry. In spite of sophisticated reliability calculations that said significant nuclear accidents were virtually impossible, the insurance industry still didn't like the odds. They insisted on government backing, even with escape clauses written into most homeowner insurance policies disclaiming liability for nuclear accidents. The actuarial masters at the insurance companies and the instincts of the public turned out to be smarter than the reliability experts. A string of disturbing nuclear accidents occurring in the 50s, 60s, and 70s was just prelude to March 28, 1979.

The Three Mile Island nuclear power plant near Harrisburg, Pennsylvania captured international attention when reactor Number Two suffered a partial core meltdown. The accident was the direct result of the unlikely coincidence of several procedural and equipment failures. Fortunately the power plant was designed, as are all U.S. power plants, to confine radiation accidentally released from the reactor core. Accordingly, even though 17 million curies of radioactivity breached the core, only a trifling 15 curies actually escaped into the environment, and there were no direct fatalities or injuries. (A curie is defined as the amount of radioactive material having the same number of disintegrations per second as one gram of radium.) Nevertheless, it was perhaps mere luck that the bubble of hydrogen created by the accident didn't explode and rupture the containment structure scattering the radioactive contents of the core over the Pennsylvania countryside. Reactor Number Two is a permanent loss; three years passed before it became radioactively cool enough to even begin to examine closely, and it has taken more than 10 years to clean up. The costs substantially exceeded $10 billion.

Accidents at nuclear power plants can be costly beyond imagination. Three Mile Island hinted at this. Another hint occurred at the very beginning of the nuclear power era at the Windscale nuclear power plant just north of Liverpool, England. In October, 1957 a poorly designed core temperature sensor led to a core fire that eventually released 30,000 curies of radioactivity into the environment. Two hundred square miles were lightly contaminated, and London, 250 miles away experienced 20 times normal background radiation. Official estimates figured about 33 cancer deaths would eventually be caused by the radiation release.

Further hints of the potential magnitude of a nuclear energy disaster have come out of the Soviet block where safety standards have always been notoriously lax, if not completely absent. (Once again, however, as with those instances noted in our consideration of the history of Islām, we would be ill advised to be too harsh in our criticism of others given the outrageous secret behavior of the United States Government in dealing with nuclear science and technology here.) The area in the southern Urals between Sverdlovsk and Chelyabinsk has been a kind of Oak Ridge and Hanford for the Soviets where several reactors were built early in the atomic age to produce plutonium. Among a number of serious nuclear problems in the area was one at Kyshtym caused by the explosion of improperly stored reactor waste products that contaminated several hundred square miles beyond habitability. In driving through this part of Russia today it is forbidden to leave one's car or even to stop. The accident occurred in 1957.

Chernobyl comes very close to representing a worse case scenario. On April 26, 1986 a steam explosion in the core of reactor Number Four destroyed the reactor and its confinement building. The cause this time was not equipment failure, but a long series of human errors. Thirty one deaths have been attributed to this accident with estimates of 30 to 40,000 cancer deaths in the future. Eventually more than 100,000 people were evacuated within a 19 mile radius of the plant. More than 100 million curies of radioactive material were released into the atmosphere over a period of three weeks. Background radiation levels exceeded 50,000 times normal in some areas. Eight years later the city of Pripyat with its numerous high-rise apartment buildings less than two miles from the reactor remains deserted by its 45,000 inhabitants, and will remain so into the indefinite future. The scene is reminiscent of an abandoned ghost town in the American west. Everybody is gone.

The promise of cheap nuclear energy has not been fulfilled. Recent economic analyses have suggested to the utilities companies that nuclear is just not competitive. This and the matters of reliability, radioactive waste disposal, and the potential for nuclear terrorism have essentially stopped the nuclear power program in the United States and considerably slowed it elsewhere. About the only application where it continues to be used is in propulsion systems for navy vessels such as aircraft carriers, cruisers, and submarines, though even here the accident rate is distressingly high.

The future does not look good for nuclear power unless fusion research becomes fruitful. However, unlike fission which started with the working CP-1 reactor, fusion research after 40 years has not yet successfully sustained a controlled reaction of any kind. A hypothetical fusion reactor would heat a mixture of hydrogen isotopes to temperatures comparable to those in the sun. At such temperatures the hydrogen plasma would destroy any containing vessel it touched, so it must be magnetically suspended in the center away from the vessel walls. Unfortunately, the problem of magnetic confinement has yet to be solved. If it is ever solved, a fusion reactor won't have those pesky, long lived radioactive materials associated with it that fission reactors have. Accordingly, fusion accidents won't leave the legacy that fission accidents do, nor indeed, will there be much to gain for nuclear terrorism.

A century of nuclear science and technology has seen man's understanding of the atom go from virtually nothing to the exquisitely detailed models touched on here and in Chapter 13. With the understanding of the atom and its nucleus and the laws that govern their behavior scientists have been able to control and use esoteric phenomena to further medicine's triumph over disease and premature death. Even the simplest X-ray has allowed the surgeon to attack cancers, bullet wounds, and fractures with a confidence unthinkable to our great grandparents. Techniques evolving out of nuclear science and technology such as magnetic resonance imaging (MRI) are approaching the magic capabilities of medicine seen only in the two TV *Star Trek* series. Scientists are surprisingly confident that in less than a decade MRI capabilities will permit fine detailed, live viewing of internal organ function. Nuclear weapons have changed the face of war, and indeed, if a nation has them, made its sovereignty inviolable except to the suicidal attacker. Controlled nuclear power, once touted as the solution to the world's energy problems, has lost its luster, and control itself may be illusory. Lacking perfect systems and perfect humans, the costs of the inevitable imperfections will leave us with a growing area of uninhabitable earth.

Chapter 16

Chaos & Fractals

For 400 years scientists have been increasingly satisfied with their understanding of the physical world and their ability to calculate and predict its behavior. The motions of the planets succumbed to Newton and the early astronomers. A hundred or more years later in the early 19th century the mathematician, Pierre Simon de Laplace, was so sure of the power of science that he claimed that in principle all the future could be calculated if only one knew all the initial conditions from which to start. Now, in the last decades of the 20th century, scientists are capable of breathtaking insights, and command an understanding of nature that seems almost magic to the layman, and indeed, sometimes to the scientists themselves. The preceding three chapters have illustrated our understanding of cosmic space and time, of materials that have given us the electronics revolution, and of the atomic and nuclear microworld that has given us nuclear weapons, nuclear power, and nuclear medicine.

Nevertheless, some problems in physics (we dare not even speak of biology) have remained intractable with no real progress evident after a century of effort. Turbulence is one example. Scientists have found it relatively easy to model and calculate the smooth flow of fluids over boat hulls, aircraft wings, and automobiles, but beyond a certain speed the flow breaks up and becomes turbulent with all kinds of eddies and whorls and nonrepetitive whips and splashes. The bubbling flow of the simplest water fountain in a public park, let alone the water in a tumbling mountain stream, is totally beyond the physicist's calculation. So is the smoke from a cigarette rising in dead still air in a smooth column for several inches and then spontaneously breaking into turbulent flow. Indeed, unlike most manmade objects, nature itself remains full of incalculable, poorly understood phenomena, atomic and cosmic matters notwithstanding. So, without any question, science has been astonishingly successful, but equally without question, there remains an immense amount yet to be understood. Perhaps as the future unfolds we will find that the very techniques of understanding will change. If they do, the concepts of chaos and fractals may well be the instruments of that change.

In high school many students take a course in algebra and eventually come across something called a "quadratic equation." As mathematics goes, quadratic equations are very straightforward and simple. And they've been around a long time. The earliest examples of quadratic equations are found on Babylonian cuneiform tablets that date at least to 1700 B.C.. Professors and young folks alike have been fussing with quadratic equations for 4,000 years. The equations have been thoroughly understood and have long since become as second nature to mathematicians as two plus two, and just about as trivial. But lurking in these simplest of nonlinear equations, undiscovered since the beginning of mathematics, are bizarre properties undreamed of even barely more than 20 years ago. And one of the many fascinating issues here is that many of these properties that may completely revolutionize how science looks at the world are as easily understood by the high school student as the original quadratic equation itself. The form of the equation that will interest us is called the logistic difference equation, or simply the logistic equation. It's been of particular interest to biologists trying to understand the fluctuations in animal populations under the influences of varying fertility, food supplies, weather, and predation.

We are not going to solve equations or get much into esoteric mathematical matters. But for an appreciation of the surprises discovered in the 1970s we must look at a couple of graphs and equations. For those with math anxiety, relax, it's going to be painless.

The logistic equation looks like this; $X_{next} = kX(1 - X)$. Here's what it means; it says that for a given percentage, "X", of the maximum possible population of critters (mosquitoes, flies, bugs, or many other beasts), and a constant, "k", that reflects a survival rate, there will be a new population that amounts to "X_{next}" percent of the maximum possible next generation. So this equation just tells you the size of the next generation if you know the size of this generation and "k." And if you want to follow the population of critters over a long period of time through several generations you can just apply the formula over and over again making sure that each new X_{next} found is brought back around and plugged into the formula at the X places so that the next generation's X_{next} can be computed. By keeping track of all these X_{next} you can follow how the population evolves. This process of using the answer the first time the equation is solved and putting it back into the equation on the right side and solving it again, and reinserting the new answer and solving again, etc., is called "iteration." In this and many other equations each iteration represents a step in time; that's why with many iterations of the logistic equation we are simply following the evolution in time of the particular population of critters. The initial condition, or value of X that we start with, is called the seed. Apart from ivory tower, theoretical playing around with the logistic equation where we might put in any seed just to get things going, one usually gets the seed value by going out in the field and measuring the current population and computing its fraction of the maximum possible population; thus our first X to get the thing going.

It turns out that for many values of "k," as one iterates the equation 10, 50, or 100 times, the solution slowly draws to a constant value, and thereafter, no matter how many iterations are carried out, the solution remains the same. This is called an attractor, and simply means that the population eventually stabilizes for a given "k."

There is however another interesting feature of the logistic equation; this is that the eventual value at which the solution stabilizes is independent of the seed value. Yes, irrespective of what seed value between zero and one (equivalent to 0% to 100%) you start with, after enough iterations the solution will always settle on the same number (always remembering that this is for a particular value of the constant "k.") This is kind of surprising, but it isn't new.

What happens if you go through the iteration process for one value of "k," get a stabilized solution, record it, and then increase "k" a little and start over through the whole routine again (same initial seed value, too)? Well, it turns out that you get a slightly larger value for the stabilized solution. In fact, if you keep doing this over and over again for larger and larger values of "k" you get a larger and larger stability solution - until "k" reaches a value of 3 and then we begin to get into new territory. The stability solution is not one value but two, that is, the solution alternates between two values with successive iterations. Strange; but easy to interpret. It just means that the population oscillates from year to year between the two stability solutions.

However, life is never simple; not for the critters, and not for us trying to figure out their population variations. As we continue to increase the value of "k" further, and carry out more iterations to find more stable solutions, we find that the number of stable solutions jumps from two to four to eight to sixteen, until we reach the "k" value of about 3.5699.

Welcome to chaos! With "k" values larger than 3.5699... there are no more stable, iterated solutions - at least for awhile. A plot of all these tedious iterations at larger and larger "k" values is shown in Figure 6 with "k" values horizontally along the bottom and the value of the stable solutions vertically along the left edge. For technical reasons "k" can have values only between zero and four, and of course the "X" values are between zero and one, that is, the fraction of the maximum possible population.

What does it all mean? Well, once we are in the chaotic region, that is, "k" values greater than 3.5699... (remember, "k" embodies any number of factors relating to the survival of the critters) then the population will vary from year to year in a totally erratic fashion. In fact, it is totally unpredictable unless you know the initial conditions, the "x" seed and "k" exactly - but that never happens in real life. The solution is chaotic! It is chaotic because even an infinitesimally small change in initial conditions can totally change the final result.

What we have just observed is chaos, defined as sensitive dependence on initial conditions, a phenomenon lurking unsuspectedly for 4,000 years in quadratic equations and discovered only in the 1970s by the mathematician-biologist, Robert May and named "chaos" in 1975 by James Yorke.

Why, one may ask, was chaos not discovered earlier? It's easy to guess the answer. Remember all the iterations we had to go through to get stable solutions? Dozens if not hundreds of iterations, each time tediously putting values into the same equation, solving, and doing it again, and again, and again. And then doing the whole thing over again for a different value of "k", and on and on and on. Who would bother with all this unless they already knew what was hidden in the computational tedium? Well, no one would, and no one did - until computers came along. Computers just love this kind of work; repetition is their forte, it's perfect for them, and trivial to program into them as well. So the computer allowed a kind of experimental mathematics that quickly revealed the chaotic behavior of both mathematical and real nonlinear systems. (This is the second time I've mentioned "nonlinear" so I'd better explain it. By "linear" one means a constant change in the output for a constant change in the input. An equation is linear if the ratio of the input to output is always the same. A linear equation plotted on plain graph paper is a straight line. Accordingly, something "nonlinear" will give an output without a fixed proportion to the input. A graph of a nonlinear equation is a curved line.)

Chaos is a property of nonlinear systems where the output has, "sensitive dependence on initial conditions." Both abstract mathematical systems, like the logistic equation, and real, physical systems, like fluid turbulence, can exhibit chaos, but not all nonlinear systems are chaotic.

In the old way of doing science and math, pre 1970, it was generally assumed that errors in the values used in calculation and measurement would behave in a well understood way, that is, generally speaking, proportionally small errors in give proportionally small errors out. However, after the discovery of chaotic behavior we can no longer comfortably rely on this. If the system turns out to be chaotic, then it has a sensitive dependence on initial conditions. Unless we know the initial conditions exactly, ie., perfectly, we cannot predict the final outcome, because the initial errors will grow out of proportion and eventually completely muck up the answer. (It is important to note, however, that chaotic systems are still "deterministic" and not random, that is, if one does know the initial conditions exactly, as for example when we arbitrarily define them, then the answer will always be the same. The output is strictly determined by the input. Chaos just says that you'd better know the input perfectly, or eventually your answer will be junk.)

In the new science and math, post chaos, a new appreciation and understanding of nature is developing. Scientists are recognizing that their immense success in the past is not leading to an end of science because of complete understanding, but is a stepping stone to greater understanding. In the past scientists avoided studying nonlinear systems, mostly because the equations were usually too difficult to solve. Nature was often idealized into linear approximations that could be solved. Fortunately this often worked, but, alas, too often it didn't, and they were stuck, as with the problem of turbulence.

Chaotic phenomena are being examined in virtually every field of science from the 'hard' sciences like physics, chemistry, and biology to the 'soft' sciences like human behavior. Since chaos is so new, most of the research is more of the nature of defining the question rather than solving the problem. Chaos is being studied in the areas of fluid flow in order to better understand and control it. Streamline shapes enhance the performance of everything from commercial aircraft to ships and Americas Cup racers to semi-trailer trucks to oil pipelines to exotic fountain displays in public parks. The dynamics of fluid flow, the atmosphere, are key to weather forecasting, a science any Sunday afternoon picnic plan relies on almost as much as the aviation industry. Mechanical engineers study the response of structures from bridges to buildings to understand the potential for unwanted chaotic behavior. Electrical engineers are concerned with the stability of individual circuits that might exhibit chaos as slight design changes are made or temperature or component characteristics change. Electrical systems like computers and computer networks have on occasion shown behavior suspiciously like chaos. Indeed, it was one of the criticisms of the Strategic Defense Initiative, SDI, that the computer control system was too complicated to trust. Would chaos have been lurking in such a system to confound the decisions for nuclear war? Chaos is clearly not simply an abstract concept.

Chaos has even been observed in chemical reactions. In one graphic demonstration, suspected of being chaotic, several chemicals are mixed in a beaker resulting in a red solution. It just sits there for a moment, then spontaneously turns blue. However, a few seconds later it turns red again. Before our amazement is over it turns back to blue, and then to red. The cycle continues with no help from the outside, it just alternates between red and blue.

It turns out that chaos is quite common in biological systems. The 40 year history of measles in New York City from 1928 to 1968 is best explained by chaotic models. We have seen it in insect populations modeled by the logistic equation, but it can appear in more complicated prey/predator relationships as well. The snow shoe hare/lynx population over a 200-year period shows strong evidence of chaotic variation.

In human behavior there is evidence of chaos as well. Putting aside the inevitable wag's comment that of course human behavior is pure chaos, in the slightly more rational and controlled situations of automobile driving, buying and selling commodities, and the decision to go to war or not, evidence is suggestive if not strong for chaos. What driver hasn't noticed that highway traffic flows just fine in off-peak hours but degenerates to stop-and-go during the rush hour, and how beyond a certain traffic density even a slight but abrupt change in speed will cause an accordion like pileup. Chaos? Sensitive dependence on initial conditions? Well, indeed, the differential equations that model traffic flow have been shown to be chaotic on occasion.

It is fascinating to find that the logistic equation has been relabeled and used to model the evolving behavior of commodity prices. Let "g" be the gain or profit (greed?), and let "P" be the current price and "P_{next}" be tomorrow's price. We have a logistic equation, $P_{next} = gP(1 - P)$. By now we all know what's going to happen when "g", (greed) gets too high; unpredictability, chaos, sensitive dependence on initial conditions; my broker wasn't feeling well after lunch, I'm wiped out! There is some suspicion that the last stock market crash, nominally brought on by computerized programmed trading of the big institutional investors, was in fact the manifestation of a chaotic potential in the market system.

In a 1984 article appearing in Nature, Alvin Saperstein offered the first model of how an arms race between nations could set the conditions for the unexpected outbreak of war. Once again, the model was our old familiar logistic equation, and though an admittedly gross oversimplification of reality, it does suggest the possibility that, under certain conditions, unpredictable (and unwanted) consequences could arise from seemingly trivial events; sensitive dependence on initial conditions, chaos, war.

"Everybody talks about the weather, but nobody does anything about it." That old saying is only partly true, for meteorologists have been trying to understand and model weather for at least two centuries - if we go back to Benjamin Franklin's first discovery that storms move coherently over long distances. Today, the U.S. Weather Bureau uses surface and satellite data from all over the world as inputs to extremely complicated weather equations that are solved on powerful Cray supercomputers. The net result, as everyone knows, is that you still can't depend on the weather - or the forecast - even though the scientists do their best. Why is this so? The answer to this question was one of the major inspirations for the chaos revolution.

In the early 1960s Edward Lorenz, a mathematician meteorologist at MIT was studying a simplified set of weather equations hoping to better understand how they worked. At one point he repeated a computer run but inadvertently entered a slightly different set of initial conditions. The answers he got were so different from the preceding calculation, using the same data, that he was driven to explore why. After eliminating the computer as the possible source of the difference he discovered the inadvertent error, a rounding off of the initial input data at parts per hundred thousand. Normally, this should have made little difference in the output, but with these particular equations it made a big difference. Lorenz had discovered the yet to be named chaos, sensitive dependence on initial conditions. He reported his discoveries in 1963 in the seminal paper, "Deterministic Nonperiodic Flow." but it was a decade before its importance was recognized. Among the conclusions of this paper was that accurate long term weather forecasting was highly problematic in principle. Since the weather equations were intrinsically chaotic, the only way they could give accurate forecasts was for the input data to be perfect, but it's impossible to get the necessary measurements of temperature, humidity, wind velocity, barometric pressure, etc. with perfect precision. Every measurement has some error, however small, and that error in a chaotic equation will eventually grow to wipe out all meaning in the solution. To get the long range forecast right you have to know all the input variables exactly. The term, "butterfly effect" was coined to illustrate the principle; it is entirely possible for the flapping wings of a single butterfly in the Amazonian jungle to initiate an exponentially growing cause and effect sequence that would eventually result in a Pacific Ocean typhoon. Since we can't expect the Weather Bureau to have the Lord's omniscience and watch all the world's butterflies - and everything else too, we'd best not expect too much from the nightly TV weather show.

Lorenz's simplified weather equations also exhibit another interesting quality. When they are plotted as a three dimensional graph, the graph shows surprisingly fine structure. Although the graph doesn't look anything like the plot of the logistic equation, it does share a certain complicated structure and scale independent self-similarity. What in the world are we getting into? Fractals.

If we take a careful look at the plot of the logistic equation (scientists like to call it a map, and from now on so shall we) we find that deep in the chaotic region (Fig. 7a) the solution becomes briefly nonchaotic, and there is an open band with a limited number of bridges across it. Each bridge represents a stable solution. These are surprising islands of stability in the midst of chaos (the Great Red Spot on Jupiter is thought to represent the same phenomenon). Of interest to us, however, is the visual quality of the bridge. It is very similar to the original, complete logistic map (Fig. 7b). If one were to magnify this bridge, one would find it too had a chaotic region, and that region, too, had stable bands bridged by little replicas of the whole map, and those replicas contained replicas, etc. ad infinitum. A figure with this self-similar structure, repeated at all scales is called a fractal.

The word "fractal" was coined by Benoit Mandelbrot at IBM in 1975 while studying the properties of strange geometric figures and the amazing variety of physical and social phenomena that can be represented by them. The property is scale independence, by which is meant that the variations in the subject behavior, amplitude or intensity or frequency of occurrence for example, are similar at all scales of consideration. He discovered that the apparently random price fluctuations of cotton really had a certain consistent internal structure that seemed independent of major economic events such

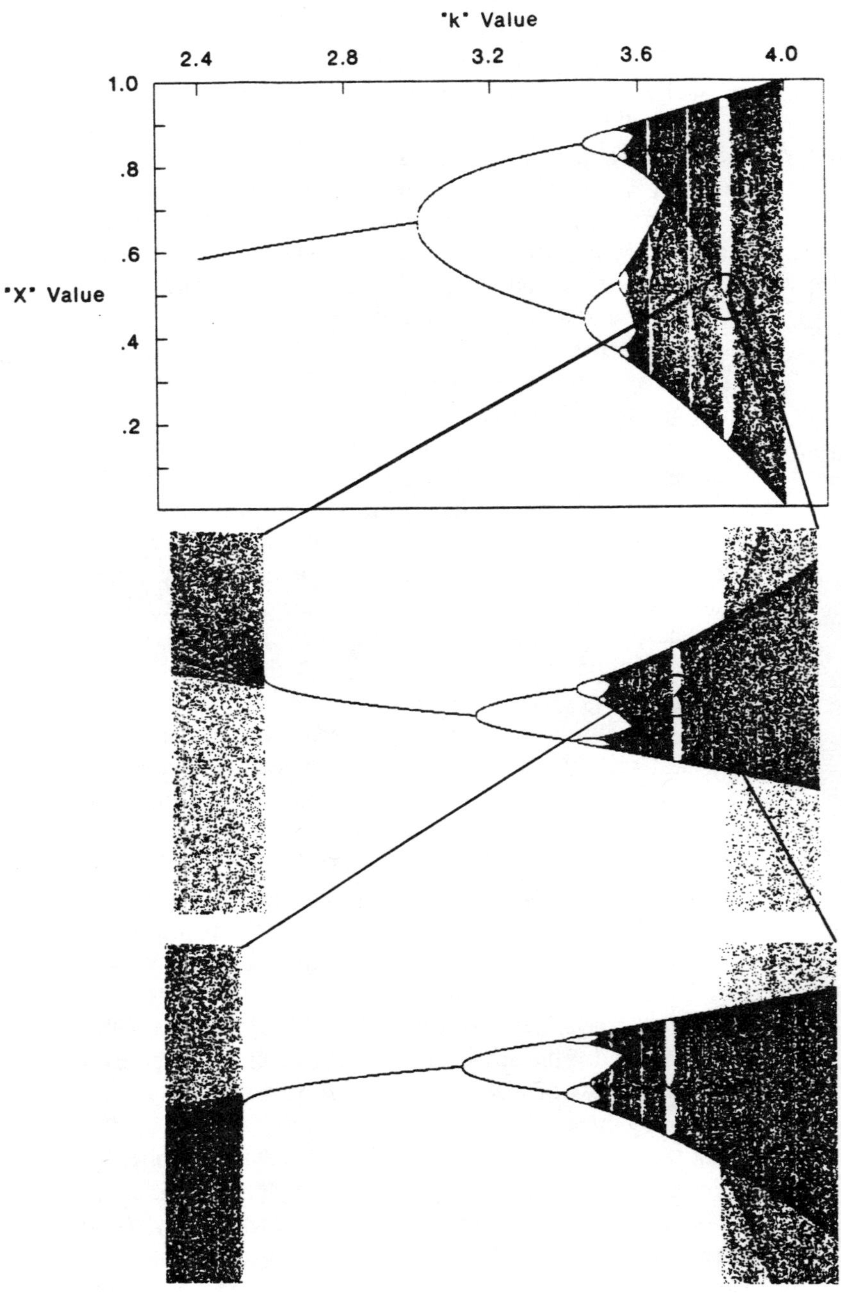

Figure 7
The Logistic Equation

as world wars and depression. He discovered similar properties in the rate of telephone line noise, in the height of the annual Nile flood, and the measured length of the coastline of Britain. What these all have in common is the way the variable in question is distributed. The variation is scale independent, and is most evident when the data are presented geometrically. It turns out that in so doing, the figures generate certain structures well known to mathematicians. These structures exhibit more than scale independent self-similarity; they have fractional dimension as well.

Almost all of us are accustomed to thinking of dimensions in terms of whole numbers, one dimensional lines, two dimensional surfaces, and three dimensional volumes. However, in the early decades of this century mathematicians developed a concept of fractional dimension which linked the intricacy of the subject structure to the measuring scale used. Accordingly, for example, when one attempts to measure the length of a coastline, the answer will depend on the length of the measuring rod used. The longer the rod, the more little twists and turns will be inadvertently jumped and thus inaccuracies introduced. Many are familiar with a similar problem when trying to measure a road distance on a highway map; we measure a different and more accurate distance as we use a smaller scale.

Intricate structures often exhibit fractional dimension, and the geometry of Benoit Mandelbrot's analyses did as well. Indeed, Mandelbrot first defined his term, "fractal", as a figure with a fractional dimension larger than its conventional dimension. Although we won't concern ourselves further with such recondite matters, it may be useful to refer to Figure 8 for some examples of regular geometric figures that have fractional dimension and are fractals.

The fractal structures of conventional geometry are usually very simple and very uniform in their scaled self-similarity. This is not the case with fractals in nature where the self-similarity is approximate, not exact, and which are so intricate as to defy representation by conventional geometric figures. Every artist knows that nature abounds with shapes radiating endless structure. The fractal geometry of nature (this is also the title of Mandelbrot's immensely influential book on the subject) is found everywhere; clouds are fractals, mountains are fractals, so are

Figure 8
Simple Geometric Fractals

river systems and coastlines, lightning bolts, trees, ferns, cauliflower, broccoli, and human bronchial and vascular systems.

As noted earlier, the logistic map, springing from a form of the quadratic equation, is a fractal. However, the quintessential geometric fractal is the celebrated Mandelbrot set (Figure 9) discovered in 1979 by Mandelbrot. Although at the time he was not concerned with the logistic equation, the Mandelbrot set is intimately connected with it. The most mathematically general form of a quadratic equation is expressed not in terms of real numbers such as 1, 2, 3, ... or 0.836, or 0.00192, or -18447, or represented by real variables such as, "X", but in terms of "complex numbers." Further explanation is not essential except to note that each complex number has two parts, a real part and an imaginary part. The rules for adding, subtracting, and multiplying complex numbers are a bit more intricate than the similar rules of ordinary arithmetic, but they are manageable. Now, it turns out that if one writes a quadratic equation in the form, $Z_{next} = Z^2 + C$, where "Z" and "C" are complex numbers, inserts suitable seed values to get things started, and iterates it just as one did with the logistic equation (remember, the solution is put back into the equation and solved again, and this cycle is repeated a number of times), then an unbelievably complicated figure can be plotted that has literally infinite intricacy, the Mandelbrot set of Figure 9.

The Mandelbrot set is a fractal object. It has self-similarity at all scales. It has a fractional dimension. It has deep and important mathematical meaning and relationship to the logistic equation, but most important for us, it is fascinatingly beautiful, and loaded with structure and figures that seem somehow familiar (Figure 10) and echoes of which precognitionally appear in the art and design of so many cultures. Is it possible that this infinitely complicated abstraction generated out of strange "complex" numbers by an equation as simple as $Z_{next} = Z^2 + C$, has somehow been anticipated by an intrinsic human sense of beauty? No one knows the answer to this yet, but some evidence is provocative. Fractal figures are in many of M.C. Escher's pictures. Parts of the Mandelbrot set immediately recall paisley patterns and Persian rug designs. Islāmic art in general is full of fractal like designs. Some southwest Native American patterns are suggestive as well as the designs found in Hmong art and American quilts.

Figure 9
The Mandelbrot Set

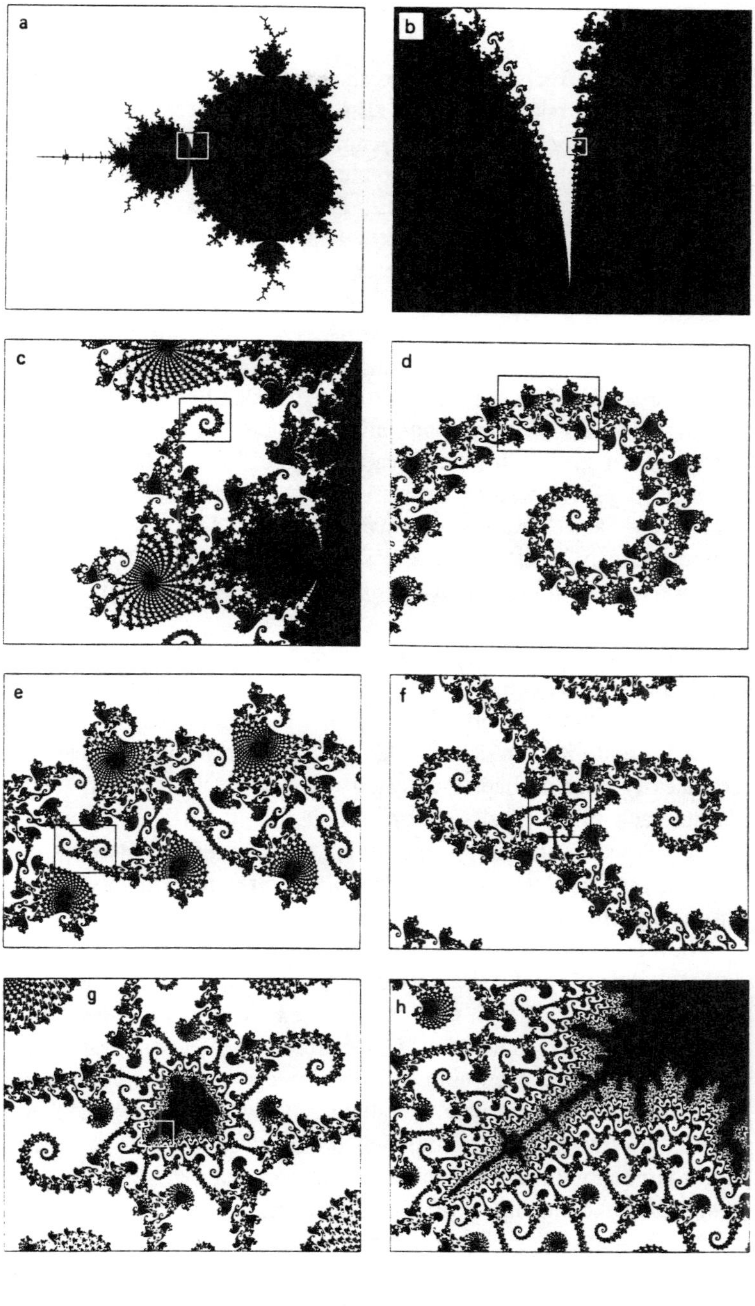

Figure 10
Fine Structure in the Mandelbrot Set

Chaos and fractal science are showing us new ways to look at and understand nature. We have discovered that many ostensibly stable systems have the intrinsic capability of falling into chaotic, unpredictable behavior. Even the "eternal," clockwork solar system appears to have this possibility. Some have pointed out that inoculations of populations against measles can have a temporary reverse effect by putting the disease cycle into a chaotic high level transient. If this ever happens, hopefully knowledge of chaotic dynamics will prevent health authorities from prematurely declaring the inoculation program a failure. We are also finding that stable behavior can exist in the midst of chaos; remember the bridges in the logistic map of insect populations (or commodity prices) and the great red spot on Jupiter? Ilya Prigogine, 1977 Nobel Laureate in chemistry, has written extensively on the interaction of order and chaos and reminds us that nature is full of so called "spontaneously" self-organized systems; the alternating color solution previously mentioned is one, and perhaps the origin of life itself is another.

Possibly the most important consequence of the discovery of chaos and fractals is the new understanding of how complexity can spring from simplicity. If relatively simple expressions like $Z_{next} = Z^2 + C$, and $X_{next} = kX(1 - X)$ can generate figures of infinite complexity, then perhaps the mystery of DNA's ability to uniquely define each of us in all of our complexity is not so difficult to accept. Physics, chemistry, and biology are currently poised at the edge of this revolutionary idea that simplicity and complexity are intimately linked, and the link goes both ways.

In a less abstract area, fractal geometry is leading to safer air travel, more realistic motion pictures, and more efficient communications. Recall that complicated natural structures like trees, mountains, and clouds are fractals. Recognizing this, technologists have been able to develop very simple equations to computer-generate extremely realistic images. These are used to mimic the environment through which the student pilot 'flies' in his flight simulator. The realism of fractal computer graphics makes a significant contribution to the effectiveness of the on ground flight simulation, giving the student a chance to cope with emergency situations under the full adrenalin rush of the real thing, but without its possible terminal consequences. Without fractal techniques, creation of realistic images would be too difficult for fast computer simulation. Similar fractal imaging techniques have been used by motion picture studios to literally create new worlds, one of the best examples being the scenes of the birth of the new planet when the "Genesis Bomb" was detonated in the *Star Trek* film *The Wrath of Khan*.

It seems that we always have more data to transmit than can be carried by our radios, telephone lines, and relay satellites in orbit. Consequently, data compression, a technique to reduce digital information to the smallest number of bits, has significant value, for example, the cost and availability of long distance phone service as well as the cost and weight of the transmitters used in deep space probes. Elements of fractal geometry have been successfully applied to this problem, particularly to image transmission which usually requires a huge amount of data. Remember that complicated fractal figures can be generated by very simple formulae, and that these figures have the property of scale-independent self-similarity. It turns out that these are the properties of parts of images in general. So it's just a matter of finding the fractal formulae to represent the parts, and use that same formula to generate similar structure at varying scales and orientations elsewhere in the picture as appropriate. A fern perfectly illustrates this concept. One transmits only a set of formulae instead of a picture, in this case the instructions for creating a single leaf, then instructions to array replicated leaves alternately along a stem, then array the stems alternately along a larger stem, and so on till the necessary detail is achieved. Indeed, the level of detail achievable is not in any way limited by the formula itself. Some fractal compression techniques can reduce the data to form an image 100 times better than conventional compression methods, and reduce transmission time by 95 percent.

The discovery and study of chaos is barely 30 years old, and of fractals, barely 20, so history is quite literally being made before our eyes. The field is so new that everywhere one looks one finds potential applications. The full magnitude of the importance of these discoveries cannot yet be guessed, but there is a growing feeling that we may be witnessing what Thomas Kuhn has called a paradigm change, a change in the very model we use to interpret nature. For hundreds of years our science has dissected nature and studied its isolated parts. The success of this technique has been phenomenal, but incomplete. Now the time may be at hand to look at the unity of nature and let the tide of knowledge draw us in a new direction.

Chapter 17

Space

The rise and fall of nations and cultures reflects deep but regional tides of history. On the other hand, science and technology, and the space enterprise in particular, will in some way or another eventually sweep up every person on the planet. By such a measure the space enterprise is arguably the most significant tide of history presented here, and perhaps even the most significant tide conceivable. Unlike political or social movements that come and go, developments in space in the 20th century signal the beginning of a tide that has no foreseeable end. It is a tide that will carry the human species from its mother planet outward into the infinite universe.

The dream of space flight goes back far beyond the invention of the rocket in 10th-century China, beyond the Greek myths to the third millennium B.C. and the Babylonian epics where a flight to Venus was attempted on the back of an eagle. The Greeks considered trips to the moon via great sailing ships, and of course we have the familiar story of Icarus and his fatal flight too near the Sun. Indeed, myths of space flight are common to many cultures. In the West dreams of space flight continued from classical times right through the 19th century with notable examples from the pens of Johannes Kepler (*Somnium*, 1634), Cyrano de Bergerac (*Voyages to the Moon and the Sun,* 1649), and Alexander Dumas (*Journey to the Moon*, 1865.)

The dream began to approach a real possibility in the 19th century as the industrial revolution and the rapidly growing power of science put at the disposal of the dreamers more and more of the necessary tools. But the time was not yet quite right, though one dreamer in particular would enormously influence the next century. In 1865 Jules Vern published a tale, *From the Earth to the Moon* which describes a voyage to the moon launched from Florida using a giant cannon to propel the lunar projectile. It so happens that this story inspired all three of the great fathers of space flight, Konstantin Tsiolkovsky, Robert Goddard, and Hermann Oberth.

Konstantin Tsiolkovsky was perhaps the most remarkable and tragic of the group for he was the first, and until recent times the least recognized of the three. Born in 1857 in an obscure town far to the east of Moscow he overcame physical handicap and professional isolation while essentially defining the modern rocket and its potential uses. His first book, *Exploration of Cosmic Space by Means of Reaction Devices* published in 1903 showed that he understood the advantages of the rocket over other methods of propulsion, that rockets were most efficient in the vacuum of space, that air caused significant resistance, that liquid fuels were superior to solid fuels, and most fundamentally, that the rocket's forward velocity was proportional to the velocity of the gases ejected out the rear. The tragedy was that practically no one outside of Russia was aware of his work, and accordingly, his genius and insights contributed little at the time to the advancement of space science.

Like Tsiolkovsky, Robert Goddard also recognized the advantages of rocket propulsion for space travel. Born in 1882 in Worcester, Massachusetts he became absorbed with space travel while still in his teens. He eventually became a professor of physics at Clark University while he continued to pursue his passion for space flight and rocket development. In 1919 the shy researcher somewhat reluctantly published through the Smithsonian Institution a lengthy article entitled, *A Method of Reaching Extreme Altitudes* where he unknowingly treated much of the same material Tsiolkovsky had 13 years before. In passing he also mentioned the possibility of reaching the moon. Within weeks the press sensationalized the idea and for a time he became known as the "Moon Man." The New York Times editorially pontificated on his foolish ideas presuming to teach him physics.[1] Naturally enough after this public ridicule he became very disinclined to announce further discoveries. Typical was his design of the world's first liquid fueled rocket and its successful launch in 1926 from a relative's farm in Auburn, Massachusetts. Though it reached an altitude of only 41 feet, it was one of the landmark achievements in the history of space science. Nevertheless, he persuaded the Smithsonian to defer publication for 10 years.

Goddard moved his research program to Roswell, New Mexico in 1930 where his work continued in comfortable obscurity. Over the next decade his experiments produced liquid fueled rockets of growing sophistication that reached 700 miles per hour and an altitude of 7,500 feet. When the Second World War came along he worked for the Navy till his death in 1945. Among the three founders of space science he was the only true experimentalist, he actually tested his theories, he actually built and flew rockets.

Hermann Oberth showed the same inspiration, imagination, and genius as Tsiolkovsky and Goddard, but whereas they were relatively obscure in their time, Oberth became well known, and subsequently had substantial impact through the development of German military rockets like the V-2 and thus the rest of 20th-century space science. Born in 1894 the young Oberth fell under the spell of Jules Vern and never looked back. With a break for

---

[1] Perhaps the most surprising element of 19th-century space fantasies is the rarity of rocket vehicles for transportation. When in earlier times all space was thought to be filled with air every mode of transportation had been imagined from swans to horses to tornadic winds to rockets. Then science showed that most of space must be a virtual vacuum, and thus sails, swans, and horses were out (rightly) as well as rockets (wrongly.) This error came from a misunderstanding of the basic physical law on which rocketry is based, Newton's Third Law of Motion which says that for every action there is an equal and opposite reaction. A rocket goes forward because gas (the hot combustion products burned in the engine) goes out the back. Tsiolkovsky, Goddard, and Oberth each independently recognized that it was only the gas rushing out the back that made the rocket go forward, and that there was no need to push against anything. Before their time, and indeed well into the 20th century, it was commonly assumed that a rocket wouldn't work in the vacuum of space since there wasn't any air to push against. Indeed, this was the error of the New York Times editorialist in considering Goddard's 'moon paper.'

service in World War One he studied physics in Rumania and Germany but failed to receive his doctorate because his dissertation on space flight was rejected. Undaunted, he had it published the next year (1923) as *The Rocket into Interplanetary Space*. Like the prescient publications of Tsiolkovsky and Goddard he independently revealed the basic concepts of rocketry, but unlike his predecessors his work was instantly recognized and became the seminal book influencing all rocket development thereafter.

By the 1930s technology was finally catching up with the millennia old dreams of space flight, but military applications took precedence as the war approached. German rocketry evolved from Oberth's inspiration into the V-2 program led by Walter Dornberger and financed by the naturally interested German Army. In 1932 the young Wernher von Braun, who cared little about military rocketry and everything about space flight, became Dornberger's first assistant. By the end of 1934 they had produced a rocket that reached an altitude of 6,500 feet demonstrating their competence and the need for a larger and more remote laboratory. In 1937 they moved from the lab outside of Berlin to a special facility constructed on a small island on the Baltic coast near the village of Peenemünde. Here the A-4 (later known as the V-2) was conceived in 1936.

When the design was finally built and launched in late 1942 the V-2 was 47-feet tall, 5.4 feet in diameter, and weighed just over 14 tons including a 2310-pound-high-explosive warhead. After two initial failures the third try succeeded; the rocket reached a speed of 3,300 miles per hour and went 60-miles high on its way 125 miles down range. Two years later it became operational, and more than 5,000 were fired at England; however the results were negligible.

With the war's end German rocket expertise was split between Soviet Russia and the US with the US getting the better of the bargain. There is a certain irony here, since both Russia and the US reaped the early benefits of rocket science from the Germans while largely ignoring their own pioneers.

After the war US rocket research progressed rather casually using a couple of hundred captured V-2s and a bevy of German rocket engineers including von Braun. For some years thereafter the US concentrated on sounding rockets. (Sounding rockets are simply shot straight up and fall back to earth after the fuel is spent.) Eventually, rocket programs were undertaken by several other nations as well including France, England, Japan, India, Israel, and China.

In the military sphere the United States and Soviet Russia began development of more powerful rockets leading directly to their arsenals of ICBMs (intercontinental ballistic missile.) In both cases their military rockets were direct decedents of the V-2. To leapfrog their own meager prewar rocket research the Russians used the few German engineers and V-2s the United States didn't get. They were also fortunate in having their own rocket genius, Sergie Korolev. It was Korolev who in the face of difficulties in developing a single large, high thrust engine, came up with the idea of clustering engines, then having clusters of clusters. The R-7, the world's first true ICBM, (in US defense parlance, the SS-6, Sapwood) had 20 engines with a total thrust of 1.1 million pounds. This cluster of clusters made it look like a slim-wasted, hoop-skirted woman with a 34 foot diameter hemline. Other, more efficient and advanced rockets were later developed for military purposes, but it was the basic R-7 and its variants that put Soviet Russia into space and continued to be the launch vehicle of choice for more than 25 years.

The first big rocket designed by the Americans was the Redstone. It evolved into the nearly identical Jupiter with a single engine of 150,000 pounds thrust. However, it was the Atlas with three engines totaling 330,000 pounds of thrust that became the first US ICBM. Shortly thereafter the Titan became the second US ICBM with two first stage engines totaling 300,000 pounds of thrust. The US, like the Soviets, developed several other military rockets, but they weren't directly relevant to the space program. Thus stood the Russian and American rocket programs in the last half of the 1950s, at the very threshold of the space age.

It was the International Geophysical Year (IGY) in 1957-1958 that inspired the first practical reach for space. In January 1955 the Soviets mentioned a general intention to put a satellite in orbit, but the US largely ignored them believing they were incapable of such a feat. Six months later the US officially announced its Vanguard program for the IGY. The next day the Soviets made it their official goal as well.

The US Navy was given the charge to develop Vanguard, an entirely civilian rocket, that was to put a 6-inch diameter, 3-pound sphere in orbit. Von Braun with the Army wasn't allowed to try because he would have used a modified military rocket, and the US wanted the world's first space venture to be on a nonmilitary launch vehicle.

However, the space age began not with a smug US walkaway, but with the Sputnik 1 launch on Oct 4, 1957. Sputnik 2 carrying a dog was launched a month later. A month after that the first attempted orbital launch of America's Vanguard ignominiously crumpling into a fireball at the moment of liftoff, live on TV.

Von Braun at the Army Ballistic Missile Agency then got the go ahead to try with a modified Jupiter-C and succeeded in putting Explorer 1 into orbit on January 31, 1958, no doubt to the chagrin of the Navy and a sigh of relief from the American people. A Vanguard rocket finally put a payload into orbit on March 17, 1958. The space age was off with a roar as its first six months saw five successes out of nine launch attempts.

An interesting, almost surreal, impression gripped the world at the time. No one expected the Soviets to be first into space. It was believed, and rightly so, that the Soviet Union wasn't technologically comparable to the United States. When the Soviets succeeded the shock created a paranoia and self doubt that totally clouded rational judgement, even in the face of cold facts. For example, a "first" by a margin of 17 weeks in an enterprise as substantial as the space program is meaningless as such, and it certainly doesn't support any notion of being "ahead." Second, the effort and will to achieve the "first" are worthwhile to note, but they reflect very little on the intrinsic technological capability of the society. As the Russian historian, Roy Medvedev has noted, "We put everything into rockets," but he neglected to say that having put everything into rockets there was virtually nothing left to make the rockets useful beyond themselves, i.e., the "first" of Sputnik 1, weighing 184 pounds and 23 inches in diameter, did nothing but radio back its presence in space. The Explorer launched 120 days later, a lightweight by comparison at 31 pounds and 6 by 80 inches, used its limited resources with relatively great sophistication and discovered the Van Allen radiation belts girding the Earth. This example at the very beginning of the space age has been repeated in kind many times over right to the present, a period now approaching 40 years. Russia has been impressive not only in its "firsts" but in its ability to more or less keep up with the US, at least superficially, all the while exhibiting an astounding lack of scientific and technological discovery. (The detailed reasons for this have been treated in Part II.)

Moreover, only two of the five successes in nine attempts during the opening six months of the space age were Russian. Though everybody was failing, the Russians were doing it a higher percentage of the time and with vastly simpler craft. In the first 39 months after Sputnik (1957-1960) the US made 66 launch attempts with 34 successes for a 52 percent success rate. In the same period the Soviets made 19 launch attempts with 7 successes for a 37 percent rate. Careful analysis of the respective technological communities shows that there was never the possibility of a meaningful space race between the Soviet Union and the United States; in the beginning the Soviet leadership were deluding themselves, and throughout the era the metaphorical reality was a track, a thoroughbred stallion, and an old draft horse disguised as a racer; Potemkin's village all over again. Nevertheless, for the remainder of this chapter I shall use achievement dates and other data not by way of comparison between the participants in the space enterprise, but merely to show mankind's progress into space.

This progress, first in unmanned space flight, followed a sequence more or less like this; earth orbit, lunar excursion, exploration of the near planets, Venus, Mercury, and Mars, exploration of the outer planets, Jupiter, Saturn, Uranus, and Neptune, and exploration of comets and asteroids. During the entire period earth orbit missions continued as the earth below and the heavens above were studied and a growing number of services such as communications and weather forecasting were provided. Meanwhile the almost independent enterprise of manned space flight was also undertaken with the sequence rather like this; man in earth orbit for short periods, man to the moon, and man in orbit for long periods. Notable equipment developments in the 1980s and 90s were the US space shuttle system and the Russian orbital space station.

Unmanned space flight evolved with amazing rapidity after Sputnik in October 1957. In the next nine years the US and USSR had 495 successful unmanned launches between them. By the end of 1991 the count was at 3123 with another 277 among eight other nations.

In 1958 there were six successful launches into earth orbit mostly for developing space capabilities and making basic scientific studies of the near earth environment. One 1958 satellite, Score, was notable as the first communications satellite,

The pace picked up in 1959 with eight successful orbital launches on various science and technology missions, as well as four trips to the Moon and one in solar orbit. Luna 1 passed within 3,600 miles of the Moon, and Pioneer 4 made it, but passed at the considerably greater distance of 37,300 miles. Luna 2 actually hit the Moon, and Luna 3 gave us our first photographs of the never before seen lunar back side. Pioneer 5 was put in solar orbit to gather solar system data. There were however a total of 28 mission failures in 1958 and 1959. With barely a combined 40 percent success rate the Americans and Russians clearly had a way to go as far as reliability was concerned. However, the early 1960s was a transition period for reliability with a mission success rate rising to a combined 90 percent in 1965 and averaging around 95 percent thereafter.

The 1960s witnessed several vigorous programs to explore the Moon and nearby planets. The motivation for unmanned lunar exploration was mostly to prepare for the expected manned missions. The Soviets, however, later changed to unmanned, science missions when they saw there was no chance of beating the US. Between 1964 and 1976 each country launched 12 successful probes to the Moon. The series began in 1964 with three Ranger missions that returned closeup pictures of the lunar surface before impact. A year later Zond 3 made a lunar flyby also returning pictures. Luna 9 then made the first soft landing and sent back pictures from the surface. There followed nine Surveyor and Lunar Orbiter missions in 1966 and 67 to map the Moon's surface and select landing sites. The last Surveyor in November 1967 was relaunched from the lunar surface and successfully relanded elsewhere. This was the last US unmanned lunar mission for more than 27 years, although the epic of manned missions was still to come.

Five more Zond flights were made between 1968 and 1970 in obvious preparation for a Soviet man on the moon mission, but nothing ever developed from them. There immediately followed five additional Luna missions that soft landed between 1970 and 1976; two included the Lunokhod vehicle that could rove over the lunar surface, and three were robot samplers that successfully returned moon rocks to earth. The last, Luna 24, completed its tasks in August 1976 ending the first lunar era.

Even at the height of unmanned lunar activity far more ambitious and sustained programs were in the works to explore the planets. In a mere 15 years beginning in 1962 successful missions were launched that visited every planet in the solar system but Pluto. A total of 29 missions have been made to Venus, Mars, and Mercury, 13 by the US and 16 by the Soviets. They began with Venus and Mars flybys made with Mariner 2 in 1962 and Mariner 4 in 1964. Two more flybys of Venus occurred in 1967, one each by the Soviets and the US. There followed a succession of more and more sophisticated missions by Venera and Mariner space craft that included orbiting the target planets and soft landing on them. The most recent Venus mission was the US Magellan which went into orbit in October 1990 and completed radar mapping its surface in 1992. The only mission to Mercury was Mariner 10 launched in 1973. It revealed a planet heavily pockmarked with impact craters and looking very much like the Earth's Moon sans dark areas (the maria).

Venus is a particularly difficult challenge for soft landing, or indeed penetrating the atmosphere at all, for the atmosphere is extremely hot, dense, corrosive, and turbulent. Between 1970 and 1985 ten soft landings were achieved by the Soviets where they discovered that the clouds perpetually shrouding the planet are composed of sulfuric acid droplets. The top of the cloud layer is 42 miles high and the bottom 19 miles. Underneath, the atmosphere appears to be clear and composed largely of carbon dioxide, however the surface temperature is near 900°F and the pressure 90 times that on earth. Venera photographs reveal a scorched terrain with thin slabs of stone like slate sidewalk tiles scattered about awaiting some Venusian denizen's weekend project.

The orange-yellow light hints at the heat and acid world in which it is enveloped.

Where Venus has held a special fascination for the Soviets, the US has largely directed its attention to Mars and the outer planets. Four Mariner and two Viking missions to Mars were launched between 1964 and 1975. The first three were Mars flybys, then an orbiter and finally the two Viking soft landers. These instrumented voyages have shown that the century long belief in a network of canals, possibly constructed by intelligent life, is an illusion. Further searches for life of any kind, including microbial, have so far yielded only negative results. Probably a definitive search for Martian life will have to await manned Mars missions that will inevitably occur in the 21st century. Mars is attractive, because, unlike Venus, it offers the possibility of successful human exploration. It is a cold planet, but not prohibitively so, rather like Antarctica. The surface with stones of varying sizes lying around on the sand reminds one of some desert regions in Arabia, except that here everything has a red hue from the high iron content. The atmosphere is more than 95 percent carbon dioxide and very thin, only 1/150th of that at sea level on earth, or about the same as that at a height of 130,000 feet, reached only by a couple of research aircraft and balloons. Traces of water ice have been found, and gross surface features visible from orbit strongly suggest that at some time in the distant past large volumes of liquid water once flowed. Altogether, Mars remains a tantalizing planet.

However, for truly mind-boggling accomplishment one must turn to the exploration of the outer planets. This enterprise is notable for two reasons, the staggering distances involved, and the unusually high success rate of the US attempts (the Soviets have made no attempts.) Not entirely facetiously, one can argue that the success rate is substantially greater than 100 percent. Five launch attempts have been made and five have succeeded. But more than that, the Pioneer and Voyager space craft continued to operate substantially beyond their design lifetime. As for the distances involved, consider this, the distance to the nearest planet, Venus, is 26 million miles. The next nearest planets to Earth are Mars outward from us and Mercury in near the Sun at 49 and 56 million miles respectively. The nearest of the outer planets is Jupiter at 385 million miles. In multiples of distances to Venus, Jupiter is 15 times, Saturn is 30 times, Uranus is 64 times and Neptune is 104 times the distance to Venus.

Pioneer 10 was launched in 1972 and Pioneer 11 in 1973. Each reached Jupiter in a little under two years. Pioneer 10 was the first to give us closeups of Jupiter and some of its moons in December 1973. After that it continued on towards interstellar space crossing Pluto's orbit a decade later (although not encountering Pluto itself.) By 1995, 23 years after launch, it was more than 3 billion miles from Earth coasting on its lonely way to the stars with a small gold plated plaque attached inscribed with the image of a man and woman, a diagram of the solar system and a line showing the route of Pioneer 10 through it, and some data to locate the Earth in our galaxy. Though a radio signal from it now takes 14 hours to reach Earth we are still in communication, and likely will continue to be till the end of the millennium. Pioneer 11 also went to Jupiter, and passed close enough to use Jupiter's gravity as a kind of whip to snap it on to Saturn which it reached six and a half years after launch. Thereafter it headed out of the solar system in the opposite direction of Pioneer 10. Both Pioneers returned pictures of modest quality yet far more revealing than any Earth-based telescopic images. Both provided gravitational data and made magnetic field measurements in the vicinity of the target planets, but both probes were lightly instrumented and rather more like scouts for the much more sophisticated Voyager craft to follow.

The Voyager program was a technological tour de force and the results showed it. The power and sophistication of the on board computing systems more than once saved the missions after their launches in the late summer of 1977. Voyager 1 was launched second, but would reach Jupiter first because of its more direct flight path.

Its flyby of Jupiter in early 1979 is one of a handful of moments in the space program that define mankind's emergence beyond his home planet. Before Pioneers 10 and 11 went to Jupiter in the early 70s our only view of the planet was through a telescope. Even the best pictures were of a fuzzy ball with vague bands and a somewhat ill defined "Great Red Spot." The Pioneers changed this considerably. The bands and spot were now much better defined and there were hints of considerable structure in the bands. Nevertheless, the pictures remained rather fuzzy as if shot through a gauze screen. Matters improved dramatically with the Voyager missions. At the time of the Voyager 1 encounter with Jupiter, I happened to be visiting the planetary imaging section at the Jet Propulsion Laboratory in Pasadena California where the space craft had been designed. Everyone knew that the returning pictures would be special; large color monitors were everywhere so that the entire staff from secretary to scientist could see them as they came in. No one was disappointed. Here we were on Earth watching breathtakingly detailed views appear line by line, live from Jupiter space. In some of the frames one could clearly see the small orbs of the moons Io and Europa floating in front of their massive master, while below them the planet swirled in small eddies of orange and brown and white. The resolution was so good that one had the feeling of being there, and I thought that *2001: Space Odyssey* would surely come true. Twenty months later Voyager 1 passed Saturn revealing detail of the planet never before suspected including the incredibly fine structure of its rings. Then it was whipped out of the solar system by the big planet's gravity.

Voyager 2's mission was a true epic. It was not simply insurance in case of a Voyager 1 failure, it was a carefully planned sister mission that would use an unusual alignment of the outer planets occurring only once in every 175 years. The planetary positions were just right to have Voyager 2 gravitationally pulled along and whipped outward successively from Jupiter and Saturn so that it would reach Uranus and Neptune. This four planet "Grand Tour" was "grand" not only for the scope and distances, but for the times; two years to reach Jupiter, four years to reach Saturn, eight and a half to Uranus, and twelve years to Neptune. The mission required extremely advanced and robust systems, every bit of computer sophistication available, and a global communications network here on Earth up to the challenge of receiving radio signals from billions of miles away originating from a transmitter having only 1 percent of the power of a microwave oven. In the end not only were Uranus and Neptune examined close up and in detail, but so were the attendant satellites of all four outer planets.

Before the space age began, ground based astronomy had found a total of 30 natural satellites in the solar system. By 1990 there were twice that many known, most discovered with the Pioneer and Voyager probes. The Voyager images of the already known satellites proved to be the greatest surprise of the program. Since Galileo's time little was known of these pinpoints of light hovering about their parent planets except their mass. All of a sudden we had in hand an astonishing set of images revealing undreamed of variety. There was Jupiter's Io, its sulphur surface gleaming orange and slick, and live volcanoes belching showers of brimstone. Then there was sister Europa with strange veins weaving across its surface making the orb look like part of a giant, bloodshot eye. Miranda at Uranus was familiarly cratered in parts, but elsewhere it looked as if someone had smeared the surface with peanut butter leaving only the stroke-marks with angular borders where they ended. Every moon was different. Every moon a new adventure for the mind. It was magnificent, and it was a magnificent achievement.

In 1985/86 Halley's Comet returned on its periodic 76-year orbit. The brilliant display of 1910 wasn't repeated, but the major space faring nations (except the US) were now ready to go out and meet it. One European, two Japanese, and two Soviet probes succeeded in showing us that Halley was a five by nine mile potato shaped piece of dust covered ice spewing out vapor from its hot spots like a boiling tea pot.

The failures of Russian and US Mars probes in 1993 created a significant setback for Mars exploration, but eventually the program recovered. Additional impetus occurred in 1996 when evidence for primitive life was found in meteoric rock from Mars. At present several nations are planning a total of 12 missions to Mars over the next four or five years.

Other recently launched or planned unmanned missions include Clementine to the Moon, Galileo to Jupiter, SOHO and Ulysses to observe the Sun, the Lunar Prospector to the Moon, and the Cassini to Saturn. There is also some planning for a Pluto mission as well as a return to Mercury sometime in the next decade or two.

Galileo was launched in late 1989 and went into orbit around Jupiter in late 1995. The trip took six years because of the highly unusual trajectory forced upon the mission by nine years of launch delays mostly due to political indecision. Galileo was launched from the space shuttle in Earth orbit. But it was sent away from

Jupiter and inward toward the Sun. Then it was picked up by the gravitational field of Venus and flung back toward the Earth where it was gravitationally whipped outward in a long elliptic orbit. It once again swung back near the Earth only to receive a final gravity kick to Jupiter. Along the way it took pictures of the astroid Gaspara, a 6 x 6 x 12 mile rock with well rounded edges and its own collection of little craters. Its nonspherical shape suggests that it might have been knocked off of something larger some time in the distant past. Before Galileo dropped into its Jovian orbit it released a probe that descended into the Jovian clouds giving us our first sample of the atmosphere.

Relatively vigorous space activity continues with manned and unmanned Earth orbit programs. Among the most notable of these are the weather satellites whose transmissions of continent wide cloud swirls we all view daily on the TV news and weather, the communications satellites that are routinely used in international telephone and TV broadcasts, and the scientific marvel of the moment, the Hubble space telescope. This eight-foot diameter telescope is weekly providing images that are rapidly shaping and securing our understanding of the near and far reaches of the Universe.

Man in space started with Yuri Gargarin's orbital flight in April 1961. There very quickly developed a competition between the Soviets and the US driven largely by national egos. The goal, explicit for the Americans but never announced by the Soviets though it was clearly evident, was the Moon. For four years progress was about even between them; man in orbit for longer and longer times up to several days, space capsules enlarged for two and three man missions, extravehicular activity, and in orbit rendezvous with other space vehicles. By the mid 60s the missions were becoming so demanding that the Soviets were unable to keep up as the US progressed on with its Apollo program to put men on the Moon by the end of the decade.

In December of 1968 an Apollo craft with a three man crew was launched from Cape Kennedy, circled the Moon, and returned to Earth. After two more test flights Apollo 11 went to the Moon and on July 20, 1969 Neil Armstrong set his foot on the dusty lunar surface. It was indeed as he said at the moment, "...one small step for (a) man, one giant leap for mankind." Two and a half hours later, after collecting some lunar samples and setting up some simple experiments, Armstrong and his copilot Edwin Aldrin returned to the Lunar Module (LM) to prepare for the return home to Earth. Before liftoff a TV camera allowed us here on Earth to look from the window to the ground around the foot of the LM. The terrain was as it had been for billions of years, hills, craters, rocks, and flat ground covered with a three inch layer of grey dust, but now with a new feature there in the foreground, the footprints of Man. One dreams at this moment and imagines a similar instant, 400 million years ago when the first amphibian crawled from the water onto a new world and left its mark in the sand.

Five more lunar landings were made with the last in December of 1972. Thereafter, manned space ventures have been confined to Earth orbit. The Soviets decided to concentrate their energies on developing an orbital space station and have been quite successful in assembling larger and larger arrays. Throughout the 1970s, 80s, and 90s Russians have acquired immense experience in long duration space flight, some lasting more than a year. The US chose to develop a space shuttle system that has many reusable parts including the delta-winged Orbiter that returns to Earth and lands like an aircraft. The Orbiter has a normal crew compliment of seven and can stay up for 10 days or more. Able to launch satellites itself, it also has served as a research and satellite service platform. Since the first orbital flight in 1981 there have been more than 70 trips into orbit and one failed mission, the catastrophic Challenger explosion in 1986. In 1995 the Russians and Americans began to coordinate their men-in-space activities with the aim of developing an international space station. In mid year the American Space Shuttle orbiter docked for several days with the Russian Mir space station for the first time.

At the moment (1996) the space tide of history has spent its first burst of energy and is consolidating itself for the next push which will surely include a large, permanently manned space station in Earth orbit. Also as surely, there will follow a return to the Moon with a permanent base there as well. Indeed, there has been considerable investigation of the practical aspects of building and living in a lunar base. In addition to periodic symposia on the subject sponsored by NASA, large US and Japanese corporations have purchased considerable quantities of lunar soil simulant (which comes from a stone quarry on the edge of Duluth, Minnesota) to explore the structural capabilities of the Moon's soil. At the University of North Dakota other researchers are working on methods to make water from lunar minerals. Of course the next big step comparable to the first Moon landing will be humans landing on Mars. It is a mission easily within our technical capability right now, but the will is momentarily absent. This will inevitably change, and sometime in the next two or three or four decades the tide of human history will burst forth from Earth and begin to encompass the planets, the stars, and the galaxy. Stanley Kubrick's Star Child in *2001: Space Odyssey* is nearing the end of its gestation.

Chapter 18

The Tides of History

Years ago, when I was working as a Soviet threat analyst for a defense company, I made it a point to cut out and save from the daily newspaper the headlines related to my chosen subject. After several months a pattern began to emerge, a kind of direction of events. At that particular time the direction was skewed somewhat from the more historic tide of the past 70 years, so I was fascinated by the question of what was going on. After another year the direction of the tide was becoming more clear; only the depth remained to be understood, and time would eventually provide the answer to that question as well. I began cutting those headlines on November 11, 1982, the day after Leonid Brezhnev's death. In less than 30 months two more Soviet General Secretaries had died and a fourth was on the scene. The direction charted by the headlines was unmistakable.

So too with events in the Middle East in recent years, the daily headlines dart in all directions, but there is an unmistakable tide flowing, albeit with fits and starts. Israel attacks Egypt in one decade, Egypt attacks Israel in another, each swears it is absolutely right and will never back down, Anwar Sadat says he would meet with Menahem Begin, they meet, the rest of the Muslim Middle East swears he's a traitor, PLO ferment ebbs and flows, Yasser Arafat declares Palestinians will never rest till the state of Israel is annihilated, Syria, Iraq, Jordan, and Saudi Arabia support him, more war, the Intifada uprising, the U.S. tries to get Israel and the PLO to talk, Israel declares them to be outlaws and terrorists, won't talk, talks with Jordanian officials, meet with PLO representatives, talks go nowhere, secret talks, public no compromise stand, Arafat changes stand, more riots, Israelis killed, Palestinians killed, talks called for, Arafat and Rabin sign limited accord, more riots, Israeli conservatives swear never, talks with Syria, talks with Jordan, etc., etc., etc.

We are awash in events. The storms of conflict seem endless. Yet, look at what has happened, and how matters have progressed over the decades. The history passing before our eyes comes into much better focus as we draw back. A direction begins to appear. If we put this movement into the context of the past 1,300 years, we begin to see hints of something like a reaccommodation with a "People of the Book." Are we beginning to sense a significant tide of history?

There is always more than one current in historic movement; the challenge is identifying the currents and understanding how they contribute to the tide. The Arab/Israeli confrontation for most of this century certainly gets most of the headlines, but is, arguably, not the most important matter affecting Middle Eastern Muslims, nor perhaps, for the rest of the world either. Long before there was a Zionist movement in history Middle Eastern Muslims for 1,000 years enjoyed the benefits of conquest, imperial power, and high culture. The historic flowering of the West beginning in the 15th century gradually replaced Ottoman and Safavid hegemony from the Mediterranean to the Caspian. Some of the early but lasting tools for this displacement were the various Capitulations noted in Chapter 3. These capitulations typify the East/West encounter; economic power driving national arrogance with the shadow of the sword growing ever greater, and an overt disregard for local custom as well as the seduction of materialism that came later. By the 1920s the decline and subjugation of the Middle East were virtually complete, an exceedingly bitter pill for peoples accustomed to think of themselves as the pinnacle of civilization. Nevertheless, even in the midst of these final humiliations a stirring of self renewal was developing. The various national movements were successful first steps for Muslims in freeing themselves of the Western yoke. The remaining task as they see it, which we continue to witness today, is to become free of Western economic control and cultural influence. Socialist and nationalist solutions have been tried with mixed results. The frustrations springing from the failures, driven by a very long history of Islāmic revivalism, have provided a naturally hospitable environment for the current fundamentalist tendencies that so worry the West. The modern challenge, however, is that the old, rigid solutions of pure Islām don't work. So we are witnessing a thrashing around for something that will work. No one can predict what it will be like, or how successful, but it is interesting to watch.

Once again headlines of recent decades suggest some direction; Western political control is discarded, nationalists and socialists governments appear and disappear, leaders are assassinated or overthrown by military coup, innumerable pledges of friendship and cooperation between Arab states, many local conflicts, little coherent action by Arab states, industries are nationalized, a greater share of oil profits is demanded and taken, economies rise and fall, political and economic reforms are periodically proclaimed, fundamentalists protest, they grow in numbers, grow much more in influence, governments suppress them, they take to terrorism, more instability, poorer economy, more discontent, fundamentalist successes - and failures, oil is pumped and squandered as nations maneuver for advantage, the primacy of religion is proclaimed - and ignored when convenient.

Is there any discernable tide here? Yes. In 1950 the Middle East was more or less a colony of the West to be used as it wished. Today the Middle East is politically independent and economically powerful, even if the economic power is often severely muted by discord and misgovernment. In 1950 a Middle Eastern Muslim was humiliated, impotent, and frustrated. Today he is less humiliated, less impotent, and angry. And with his growing sense of dignity and power he is becoming a bit more accommodating and less defensive in some areas, witness the slow but sure rapprochement with Israel. The tide seems to be in the direction of reestablishing a Muslim identity and global significance. Whether this will mature into an historic tide or be washed out by other influences will not be evident for decades or a century or more. What is clear is that the reemergence of the Islāmic Middle East in world affairs is one of the defining tides of the 20th century.

Now consider the tide of Russian history and its 20th-century blossoming in the Soviet state. In the mid ninth century the East

Slavs who were in a state of constant external and internal strife fell under the rule of Viking traders and succumbed so thoroughly that they have been known ever after by the name of their rulers, "Rus." But what mark did the Vikings leave besides a few Slavicized names (Oleg = Helgi , Olga = Helga, Igor = Ingwarr, and Vladimir = Waldemar) and more political unity than before? Very little; after a century the Norse influences had vanished, and the Rus were thoroughly Slavic. A century after that the Rus began their long conversion to Orthodox Christianity, but *they* were the ones to choose, and from the many possibilities they chose a religion very congenial to their already existing Slavic world views of the inevitability and efficacy of suffering. Next, the Mongols came and ruled them for 240 years. The net result was more suffering followed by new and disturbing contacts with a West that offered many desirable things, but was too assertive and vigorous for comfort. The stress of the Tatar Yoke and the refuge of the Orthodox Church helped form a special self image, so when Constantinople fell to the Muslims in 1453 Russians saw themselves as the leaders of the last spiritual hope of the world, the Third Rome. In a couple of centuries Peter the Great threw the door open to Empire and Catherine the Great consolidated it. Alexander I swaggered with Napoleon, and triumphed over him. Russia seemed to be a European power of the first magnitude. French was spoken in the drawing rooms, and the aristocracy forgot its Russian.

Yet, underneath it all the vastness of Russia hid tens of millions of peasants, more than 90 percent of the population, who continued to suffer and worship as their Slavic ancestors had for 1,000 years. By mid 19th century the Western educated intelligentsia was becoming acutely aware of its distance from the soul of the peasantry, and thus began a fervent search for its Slavic roots. While the educated anguished, the peasantry began to evolve and surrender its village culture for sterile urban factories and homes. The spiritual vacuum that followed was filled by Lenin (and Stalin) who provided all the suffering, all the messianism, and all the semi-divine images of a protective father figure that the Russian people could possibly want. So even in the quintessential Western type of factory, even sans Christian religion, Slavic needs were being fulfilled. The substance was there, only the name attached to it was changed. For 70 years this situation was continued and consolidated, but in the last analysis Russian socialism couldn't compete in the Western world it chose to challenge, and it collapsed. So what's left? Peasant or middle class sophisticate? Atheist or Russian Orthodox Christian? Slav or European?

It is striking to see the continuity here. The tide of Russian history has been the very slow transformation of the pagan, Slavic, Polyanian and his brethren through a millennium of backwardness and suffering into a people who put the first satellite in orbit, then the first man in orbit, and who were essentially the equals of the Western best in nuclear weapons, aircraft, tanks, and warships. The continuity is most manifest, however, in the suffering and apparent elemental detachment from the comforts of life. Thus, with all of the modern accoutrements of the West ostentatiously flaunted, there has been in fact an undercurrent of the old ways, manifest in part by a crazy quilt economy, alcoholism, and complete indifference to the quality of what is produced. Russia's high achievements have been at enormous cost, because they haven't come naturally as in the West. Nevertheless, the effort and experience have left their mark. Though it is possible that Russia may slip to the levels of relative backwardness it has experienced in the past, it's highly improbable that the people will accept it for long.

The problem at hand for Russia is how to comfortably reconcile her spiritual needs springing from 1,000 years or more of Slavic ethnicity and Orthodox Christianity with Western materialism without succumbing to materialism's corruptiveness. If they can do that, they will not only save themselves but the West as well.

It is interesting to note that both the Islāmic Middle East and Orthodox Russia are striving for the same thing, a way to satisfactorily integrate their own long traditions with the best of what the West has to offer. The challenge is how to do it without also being corrupted by the seamier elements of Western culture, and without sacrificing the unique spiritual qualities that have made each of them great. There is now and in the coming decades an historic opportunity for them both. One hopes it is not squandered.

Unlike the tide of Muslim reemergence or Russian self discovery, the tide of science and technology is in no way problematic. The 20th-century scientific tide of history now flowing before our eyes is the most profound, universal, and influential of any in history. Very few people on earth have escaped it. One is reminded of the cargo cults in the western Pacific after World War II. Primitive tribes sometimes constructed wooden likenesses of aircraft and control towers hoping thereby to entice desirable goods from the sky as they had seen happen during the brief sojourn there of Allied air forces during the war.

Nuclear science and technology may represent the first essentially hubristic creation of the human species. This product of the mind, this science, has now made it possible to literally destroy the entire world, and to do so with relative ease. This sobering capability is not yet fully appreciated by some who would join the nuclear club. Nations with nuclear weapons, even very small, weak nations, can now cause the strongest to consider very carefully any plans for physical dominance. Sovereignty-threatening war among nuclear powers is suicidal. This has radically transformed the nature of war, a nature that has remained relatively static from the dawn of history until mid 20th century. However, the potential for nuclear mischief resides not only with weapons and warfare. Three Mile Island and Chernobyl only hint at the other possibilities. The disaster at Chelyabinsk is more explicit. Mutant weeds cover huge piles of topsoil scraped off the hottest areas still dangerously radioactive almost two generations after the accident. The local residents call the piles, "The graveyards of the Earth."

It is difficult to realistically imagine a future where nuclear weapons won't eventually be used, at least in regional disputes between small nations led by petty dictators. The passion, temptation, and absence of *savoir-faire* in these circumstances, added to the virtual certainty of weapons availability springing from the loss of control of such weapons as the states of the old Soviet Union continue their turbulent evolution, will surely make nuclear warheads available. Other nations such as Iraq, Iran, North Korea, Pakistan, and Argentina may choose to build their own. Once available in this milieu there is every precedent for their use, witness the readiness of minor nations to use prohibited chemical weapons in the past.

As the history of nuclear science and technology passes before our eyes we will likely see a decline in the development and use of fission power with its messy issue of environmental pollution. Fission power could be replaced by the possible emergence of fusion power as research draws us closer to feasible commercial systems. However, that eventuality appears to be some time off. In the meantime alternative energy sources such as solar may truncate the nuclear option entirely. There is no way to forecast the matter, but it will be interesting to watch it evolve, and for those so moved, to try and influence the evolution.

Nuclear medicine has been impressive in its ability to visualize internal structure and destroy rampaging cells within the human body. Alas, it has always had the corollary effect of being invasive. This is a nice way of saying that these procedures cause damage to healthy tissue at the same time that they are dealing with the pathology. This is a fundamental weakness of all such procedures, and should be avoided where suitable alternatives can be found. At the moment there are few such alternatives, but the signs exist suggesting a time will come when the medicine of diagnostic and therapeutic ionizing radiation will end. The development of magnetic resonance imaging (MRI) on the one hand and molecular biology and improved surgical techniques on the other may be the direction of this demise. These are areas where the history of this science and technology are very promising and worth watching.

The discovery of fractals and chaotic behavior appears to be setting the stage for a change in how science approaches and views problems. Beyond this, it may become a new paradigm for extension into the social sphere, how humans, especially in the West and particularly in the United States, formulate problems and define acceptable solutions. Since Newton and before, we have traditionally behaved as though all problems and their solutions could be defined. If a ship rolled over in a storm, then it must have been improperly designed. If rush hour traffic degenerates into stop-and-go progress, then it must be because there aren't enough highway lanes. If the TV weather forecaster predicts a cloudless day but we get rain, it must be because the forecaster didn't have a sufficiently powerful computer at hand. If price controls, or arms control agreements fail or in fact lead to the opposite of the desired result, then there must be dishonesty at work. The world has traditionally been viewed as if it were a machine with machine like determinism. Deterministic or not, chaos has shown us that many systems, even very simple ones, can behave in extremely complicated and unpredictable ways. Chaos is forcing us to reconsider how entire systems behave and evolve, an important point in a world increasingly driven by large computer systems and networks. Remember, it was only recently that the United States government was spending billions annually on the computer controlled SDI "Star Wars" defense. It's a sobering thought that such a system might behave in unpredictable ways. Others from time to time have considered automating other important facets of our culture. All perfectly reasonable from the Newtonian point of view, but is Newton up to it? Chaos says no. Chaos theory may be giving us a new respect for the threshold sensitivity of natural processes to major change. For example, recent discoveries of the rapid change in the black spruce tree line in Canada, and in deep ice cores from the Greenland ice cap have revealed the possibility that the onset of ice ages may have come not in millennia or centuries but in less than a decade. Major climatic change triggered by what? How sensitive is the trigger, and how close are we to that threshold? Questions like these are more than academic in the knowledge of the phenomenon of "sensitive dependence on initial conditions."

Fractal geometry represents another side of chaotic behavior. The border between basins of attraction for chaotic systems (the starting regions where a system might evolve in one way or another) is fractal. Beyond that, fractals are so natural that it's a wonder nature was ever viewed any other way, but it was (a tribute to the power of the mental paradigm of Euclid and Newton.) Indeed, Piet Mondrian, the Dutch abstract painter was so entrapped by this paradigm that he is reputed to have said that the branch of a tree is nature's inferior attempt to create a line. We now know what Mondrian didn't, that this "inferior attempt" is a hint of a breathtaking world of intertangled simplicity and complexity that many might say far surpasses the spare shapes of the cubists and their followers. Fractal geometry and its products are at once homey, natural, comfortable, fascinating, and profound. This alternative to most of 20th-century art forms has yet to be sounded, but the occasion may be imminent.

Cosmology is impacting us more indirectly and subtly than chaos, nuclear science, or electronics. Our self image is being reformed by new views of the universe and our place in it. The very concepts of time, eternity, and creation are being questioned, and naturally enough, our views of God and religion. Anxiety already appears from time to time over what would we do if we found other intelligent life in the universe. The sigh of relief was almost palpable in 1976 when the Vikings 1 & 2 Mars landers found no sighs of life. There can be little doubt that the discoveries awaiting us as the Hubble Space Telescope and the present and future COBE like satellites that map the structure of the very young universe probing the farthest reaches of time and space will profoundly transform our notions of our human self. Theoretical physics, one of the highest products of the human mind and so characteristic of the scientific revolution itself, seems destined to create some kind of theory of quantum gravity that not only will permit a unified Theory of Everything but allow mankind to take the intellectual step into the nether world of the Planck Era and right to (and through?) the beginning of time.

The growing knowledge and appreciation of the natural order of the universe will inevitably reshape our view of supernatural origins and divine influence. This will not necessarily replace religion and a belief in god, for such beliefs clearly reflect a very deep human need. What is more likely is some sort of synthesis, possibly like the deist philosophy of a creating but non-interfering god. However, no matter what the eventuality, a slow, deep change seems unavoidable as sophisticated knowledge of the physical universe filters from the specialists into the community of common humanity.

Transistorized radios are found everywhere within range of a radio station, even with the remote tribes and rural villages of Africa, South America, central Asia, and the western Pacific. Nothing of significance escapes being immediately broadcast throughout the world. The cultural impact of this unifying instrument has, ironically, released energies of political faction no one knows how to handle. The impact of the transistor has been incalculable in facilitating the communication and information revolution that is transforming the world. Individually and collectively we have virtually achieved the omniscience of gods (without, alas, an accompanying wisdom or omnipotence.)

The computer, as one example among hundreds possible, has become so powerful and flexible that it's become a metaphor for life itself. While biologists are still groping towards the fundamentals of life, computers are discussed in terms of neural networks that simulate the brain's learning processes. (Intel has made a special 'neural' chip with 3,700,000 transistors on it that can complete 10 billion operations per second.) Expert systems, artificial intelligence, and the creation of virtual reality are other products of the computer that either already exist or shortly will. The metaphor for life is completed by the new problems that computers experience when their software is maliciously "injected" with computer "viruses" that "infect" the program and prevent its successful execution. The military, and business institutions as well, are deeply concerned with electronic "germ warfare" and tampering.

The tide of electronics development continues to accelerate, and the evolution of the computer reflects the very essence of this movement. The impact of the microprocessor chip has been staggering, but so far surprisingly peripheral, though microprocessors are hidden everywhere, in our automobiles and traffic lights, our microwave ovens and other appliances, in our radios and TVs, and even in our credit cards, and of course in our home computers. Apart from our home computers, the microprocessor has been largely invisible and transparent (we don't even know its there). But when it isn't invisible, as in the infamous programmable VCR, where few of us ever get it right when we want to record some future program, it's decidedly not "user friendly." This has been the major barrier to making computers central to our common, everyday activity around the house. The home personal computer has not changed this situation, indeed, it is operationally nothing more than a super complicated VCR with no prospect of widespread utilization like the telephone, notwithstanding the hype of technocrats, Wall Street, and computer buffs.

However, this barrier is on the threshold of disappearing, for at this very moment the technology of voice recognition is appearing on the scene. Very shortly, surely by the year 2001, even the home computer will be able to respond to complex voice questions and commands with the felicity of a human being. If we can say it, the computer can understand it - and act accordingly. Voice recognition technology already exists with the implementing software for home computers priced within easy reach of personal computer owners. The technology is not very good at the moment, but that will change in the next several years. This is inevitable as the next generation of microprocessor chips appear in the 1996 to 1998 time frame thus easily making possible the necessary computational power for intelligent and speedy voice recognition. A hint of how this will facilitate life's little chores as well as profoundly affect everything we do may be seen in any TV episode of *Star Trek, the Next Generation* or in the 1969 film *2001, Space Odyssey*. We will be talking to our computers and computerized equipment just as we talk to one another and as the characters talk to Computer in *Star Trek* and HAL in *2001*.

There is an irony building here that is particularly relevant to the point of view of this book. China, a relatively backward country in recent centuries, is making a special effort to implement voice recognition for its personal computer users. They have an immense need for this, because their written language is much too cumbersome for keyboard computer input. It turns out, however, that the sounds of their spoken language are much easier for computers to recognize than English or other Western languages. Accordingly, there is every expectation that voice recognition technology will be perfected first in China and come into common use there well before it does in the West. This will facilitate computer use to unimaginable levels. Combine this with her ambition, work ethic, and one-forth of the world's population, and one cannot escape concluding that another tide of history, untreated here, may dominate the 21st century.

"Space, the final frontier: These are the voyages of the starship Enterprise." Thus began the episodes of the two *Star Trek* series. In 1966 when the broadcasts began, momentum was building for the first manned mission to the Moon. Then in 1969 the film *2001...* focussed the celebrated moment with the concluding scene of a star child embryonically floating in space before the Moon. Within days Armstrong had done it. The technological prophecy of these dramas is perhaps deeper than many appreciate. The first half of the 20th century saw the tentative development of rockets and other technologies that would enable the first ventures into space in the second half. After a burst of activity in the 60s and 70s spectacular new programs that would extend the reach of our species were curtailed. Nevertheless, the utility of Earth-orbit space had already become substantial. Indeed, it has become so great that orbital activity has begun to create a space traffic jam. Today nations are in such hot competition for favorable geosynchronous orbits that international agreements have been made to regulate matters.

Currently we are in an era of consolidation before the next big leap to Mars. The step beyond that will be transcendent. The history of our species, whose little tides we've been examining, will manifest the destiny of sentient life itself and scatter outward from our star to encounter historic tides elsewhere in the Universe. Science and technology will be the vehicle. This tide is so deep and profound that it's not possible to overestimate its consequences. The pleasure and anticipation in knowing this and thus being able to appreciate its birth and evolution are part of the payoff in watching the tides of history.

Glossary

Pronunciation Guide for Arabic Words

Short "a": between the vowel sounds of "hat" & "fur."
Long "a": (ā): like a in "hard."
Short "i": as in "pit."
Long "i": (ī): like vowel sound in "feet."
Short "u": as in "hook."
Long "u": (ů): as in "clue."
Glottal stop (') as where "t's" should be in "bottle" as it is pronounced by many New Yorkers.

Glossary

10^x notation: a convenient way to write very large or very small numbers where "x" is the number of zeros before the decimal point for numbers greater than 1 and "-x" is the number of places after the decimal point for numbers less than 1. For example; $1 = 10^0$, $10 = 10^1$, $100 = 10^2$, and $0.1 = 10^{-1}$, $.01 = 10^{-2}$, $001 = 10^{-3}$etc. So one million (1,000,000) would be represented as 10^6 and one-billionth (0.000,000,001) would be represented as 10^{-9}.

Abbāsid: as in the Abbāsid Caliphate, of or related to the house of al-Abbās.

Amir: commander, governor, or prince, usually a nonreligious leader.

Anatolia: the Asian part of what is now Turkey.

atom smasher: a machine that speeds particles such as electrons or protons to high energies so that they can be "smashed" into atoms and the reactions studied. Also used to produce radioactive elements for medical applications.

ayatollah: from the arabic, *ayat allāh*, meaning token, or sign of God; a Shī'ite religious leader.

Bahā'ī: a modern religion stemming from a 19th-century Shī'ite schism in Iran.

Balkans: countries occupying the peninsular area between the Adriatic Sea and the Black Sea; recently, Albania, Yugoslavia, Bulgaria, Rumania, Greece, and European Turkey.

Bhagavad Gītā: "Celestial Song" or "The Lord's Song"; a relatively short poem found within the Hindu epic, t h e *Mahābhārata* where the god Vishnu comes to earth as Krishna and instructs a prince on ethics and the nature of God.

binary: in mathematics and computer science a method of representing any number by a series of 1s and 0s with positional values equal to powers of 2. So the first position on the right is equal to 2^0 or 1, the second position just to the left is equal to 2^1 or 2, the third equals 2^2 or 4, the forth equals 2^3 or 8, etc. For example, the binary equivalents of the numbers 1, 2, 3, 4, 5 are 001, 010, 011, 100, and 101.

bit: a digital unit equal to either a binary "0" or a "1".

black body distribution: the distribution of radiant heat energy from an ideal radiator called a black body. At each temperature an ideal black body radiates a characteristic and unique distribution of heat energies.

BMEWS: Ballistic Missile Early Warning System; a string of radar stations stretching from Alaska to Greenland to Great Britain.

Boolean algebra: an algebra developed by George Boole in the 19th century that deals with the truth or falsity of logical statements. With computers "true" or "false" can then be represented by 1s or 0s and easily manipulated digitally.

***boyar*:** a high ranking Russian nobleman that helped run the government; title abolished by Peter the Great.

Brahe: Tycho, (1546-1601) Danish astronomer noted for his accurate measurements of planetary orbits that led Kepler to his three laws of planetary motion.

breeder reactor: a special kind of nuclear reactor that can not only produce useful power but also convert common uranium-238, through neutron capture, into plutonium-239 at a rate exceeding a reactor's consumption of plutonium.

byte: a digital unit equal to 8 bits.

caliph: from the arabic, *khalīfa,* meaning successor, in this context, successor of the Prophet (a religious/secular leader).

CAT: "Computerized Axial Tomography", a technique where successive beams of X-rays are passed through an object, each at a different angle, then the different X-ray absorptions are measured, analyzed, and used by a computer to reconstruct a detailed image of the inner structure of the object. The pictures are very much better than those obtained with conventional X-ray techniques.

cathode ray tube: a tube just like a TV picture tube, so called because in the late 19th century the electrons that flowed from one end of the tube to the other were not recognized as such and were called "cathode rays."

CD-ROM: "Compact Disk Read Only Memory", a compact disk just like those we buy at music stores except that they contain encoded information readable by a computer.

CCD: "Charge Coupled Diode (or Device)", an array of very small, independent, light sensitive semiconductors on a single chip.

Clementine: unmanned U.S. spacecraft sent to map the Moon in February 1994.

COBE: "Cosmic Background Explorer," artificial satellite with a very sensitive radio receiver to map possible variations in the universal 2.7 K microwave background.

confinement: a technique or structure to partially restrict dispersion of debris and radioactive materials if a nuclear reactor should catastrophically fail. The reactor building at Chernobyl was the only confinement in that episode.

containment: a technique or structure to totally restrict dispersion of debris and radioactive materials if a nuclear reactor should catastrophically fail. The circular, steel and concrete building at Three Mile Island is a good example of effective containment.

Copernicus: Nicolaus, (1473-1543) founder of modern astronomy who proposed the earth rotates once a day on its axis and all planets revolve in orbits around the sun.

core: the central, active part of a nuclear reactor where the nuclear fuel is usually contained in long rods inserted in a neutron slowing medium (a moderator) such as graphite or water.

Coulomb: Charles Augustin, (1736-1806) French scientist who developed the quantitative law of electric attraction and repulsion between electric charges.

curie: a measure of the amount of radioactive material; one curie is the quantity of radioactive material that has the same number of disintegrations per second as one gram of radium, 3.7×10^{10} per second.

Cyrano de Bergerac: 17th-century soldier and author; the tragedy named after him has in fact little to do with his life.

***dacha*:** in Russia, a summer or vacation home.

detente: a period in the mid 1970s of relaxed tensions between the West and Soviet Russia.

DNA: deoxyribonucleic acid; a molecule in the form of a double helix that carries genetic information.

Druze: a distinct religion with their own scripture originating in the 11th century as a small offshoot of the Ismā'īlī Shī'ites and surviving today mostly in Lebanon.

duma: originally a council of boyars, later an elective legislative body with no class membership requirements.

electron: the lightest 'elementary' particle (apart from the neutrino and the massless light photon) carries one unit of negative electric charge. Electrons are what flow in a copper wire as electric current.

electron volt: a very small but convenient unit of energy used by physicists, especially when talking about atomic and nuclear processes.

elementary particles: a term used, especially in the 1950s and 1960s, to indicate what at the time were thought to be the most basic constituents of matter, for example, protons, neutrons, and electrons.

EMP: "ElectroMagnetic Pulse", a pulse of electromagnetic energy, typically originating from a nuclear explosion very high in the earth's atmosphere, that can destroy sensitive electronic equipment at distances of thousands of miles from the blast.

Falconet: Étienne-Maurice, 18th-century French sculpture.

Faraday: Michael (1791-1867) English scientist who discovered that an electric current could be made to flow in a conductor if it was moving with respect to an enveloping magnetic field.

Fātamid: as in the Fātamid Caliphate, of or related to Fātima, the daughter of Muhammad and wife of Alī, the fourth caliph.

fission: the splitting of a heavy atom, such as uranium, into nearly equal parts.

fusion: the joining of two light atoms such as hydrogen to form a heavier atom such as helium.

Galileo: Galileo Galilei (1564-1642) Italian astronomer and physicist best known for demonstrating from the leaning tower of Pisa that all bodies, irrespective of composition, fall with the same acceleration (air resistance aside), improved the telescope, discovered the moons of Jupiter, and supported the Copernican, sun-centered view of the solar system (for which he was censured by the Catholic Church.)

Galileo: unmanned U.S. spacecraft launched to Jupiter in 1989, arrived in 1995 to probe its atmosphere and study its major moons.

gamma rays: identical to X-rays except usually of a higher energy.

Gaspara: a potato shaped asteroid about the size of the core of Halley's Comet.

geosynchronous: as in "geosynchronous orbit," meaning orbiting the Earth once a day thus in synchrony with the Earth's ratation and apparently stationary overhead.

***glasnost*:** a Russian term meaning frankness, openness; a policy urged by Gorbachev to reveal corruption and inefficiency in government.

***hadīth*:** traditions and sayings of the Prophet, also applied by the Shī'ites to the traditions of Alī.

***hajj*:** pilgrimage to Mecca

half life: the time taken for half of a radioactive material to decay; for example, one curie (3.7×10^{10} disintegrations per second) of strontium-90, 28 year half life, would still be emitting 1.85×10^{10} after 28 years, 0.925×10^{10} after another 28 years, 0.4625×10^{10} after another 28 years, etc. A rule of thumb is that it takes ten half lives for any quantity of radioactivity to become negligible.

***hanīf*:** a pre-Islāmic, pagan, Arabian monotheist given to much meditation, not Christian or Jew.

Henry: Joseph (1797-1878) American scientist discovered method of inducing electric current independently of Faraday, also made numerous other contributions to the science of electromagnetism.

***hijra* (or hegira)**: flight or withdrawal; Muhammad's emigration from Mecca to Medīna in 622.

Hubble Space Telescope: a U.S. astronomical telescope launched into Earth orbit in 1990 and repaired in 1993.

IGY: International Geophysical Year, a worldwide cooperative program in 1957 and 58, among more than 70 nations to study the Earth.

***imām*:** leader or exemplar, prayer-leader, a spiritual leader of the community; in Shī'ite Islām, the divinely inspired successor to the Prophet.

Jaxartes: now called the Syr Darya; the eastern-most of the two great rivers flowing into the Aral Sea, it originates near Tashkent in Uzbekistan.

JPL: Jet Propulsion Laboratory, operated jointly by NASA and Cal Tech for the purpose of developing civil and defense spacecraft.

Ka'bah: the central shrine of Islām, a roughly cubic building in the Grand Mosque in Mecca

Kepler: Johannes (1571-1630) German astronomer discovered three laws of planetary motion.

Khwārezmia: also Khwārezm, area just south of the Aral Sea astride the Oxus river.

Krishna: popular Hindu god and incarnation of Vishnu, one of the supreme gods.

Luna: a very extensive series of Soviet unmanned probes of the Moon from 1959 to 1976.

Lunakhod: a Soviet robotic vehicle landed on the Moon with Luna 17 & 21.

Magellan: a U.S. space probe that orbited Venus and sent back very detailed radar maps of the planet's surface.

mahdī: the awaited one or specially guided one who will be the returning "hidden Imām"; divinely inspired leader to bring perfect justice to the world just before its end.

Mamluke: military caste of slaves who ruled Egypt in medieval times.

Mariner: a series of unmanned U.S. probes to Venus, Mars, and Mercury

mass excess: light and heavy atoms have excess nuclear mass compared to those of more middle weight, the difference being manifest as binding energy which is released in fission and fusion.

mawālī: plural of *mawla;* client, helper, freed slave; non-arab Muslims in the early centuries of Islām.

Mesopotamia: the region in modern Iraq astride the Tigris and Euphrates rivers.

metropolitan: a regional head of the Eastern Orthodox Church, like a bishop or archbishop.

micron: one-millionth of a meter, or about a twenty-five thousandth of an inch (one-fiftieth the thickness of a human hair).

microwave: short wavelength (typically a few inches) radio waves often used in radar, long distance telephone communications, and electronic ovens.

Mir: a modular Soviet space station in Earth orbit since 1986 with additional sections being added from time to time.

Moors: a mixed North African (Berber) and Arab people.

MRI: Magnetic Resonance Imaging; an imaging technique that maps a distribution of matter by sensing its interaction with suitable radio and magnetic fields. It is considered "noninvasive" because it doesn't use any damaging radiation or radioactivity.

Mughul: a Muslim dynasty, originally Mongols, that ruled India from the 16th to the 18th centuries.

Musta'līs: Egyptian faction that with the Nizārīs, Iranian faction, split the Ismā'īlīs Shī'ites over the succession of a Fātamid caliph.

Nazārīs: see Musta'līs.

Nejd: the rocky, central plateau region of Arabia bounded on the north and south by the sandy Nafūd and Rub al-Khali deserts respectively.

Neptune: the eighth planet from the sun, the next beyond Uranus.

neutron: an electrically neutral 'elementary' particle usually found in atomic nuclei, has a mass slightly more than a proton.

Newton: Sir Isaac (1643-1727) English scientist, developed the calculus, discovered the three laws of motion and a law of gravity among many other achievements. The greatest and most influential individual in the history of science.

nihilist: a member of one of a number of Russian terrorist groups of the 19th and early 20th centuries who believed government had to be destroyed before reform was possible.

Ohm: Georg (1787-1854) German scientist who discovered "Ohm's Law" giving the basic relationships between electric current, voltage, and resistance.

OPEC: Organization of Petroleum Exporting Countries; organization which tries to regulate crude oil production and prices.

Ottoman: the name given to a long line of Turkish Sultans who ruled much of the Muslim world beginning in the 13th century with Osman I.

Oxus: now called the Amu Darya; the western-most river entering the Aral Sea flows 1,600 miles from the mountains of Afghanistan.

particle accelerator: a more modern term for atom smasher; a machine that speeds up particles from millions to trillions of electron volts energy.

particle physics: the branch of physics concerned with studying the basic structure of matter at the individual particle level.

patriarch: most senior ecclesiastical heads in the Eastern Orthodox Church.

perestroika: "restructuring" process of Mikhail Gorbachev to attempt reform of failed socialism in Russia.

PET: Positron Emission Tomography; very similar to CAT except that the source of the mapping radiation is an internally taken radioisotope which emits positrons, the antimatter equivalent of electrons.

Pioneer: series of U.S. unmanned probes to Jupiter, Saturn, and Venus.

Planck Era: the time before 10^{-43} seconds of age of the universe where the quantum laws of physics involving Planck's constant suggest that space and time have no meaning as we know it.

Planck Time: 10^{-43} seconds after the hypothetical beginning of the universe; the moment when gravity ceased to exceed or equal the other fundamental forces. The end of the Planck era, thereafter space and time take on characteristics familiar today.

plasma: a completely ionized gas composed of electrons and atomic nuclei.

plutonium: fissionable manmade transuranic element.

proton: positively charged, heavy 'elementary' particle, and the basic constituent of the atomic nucleus.

quantum gravity: a yet to be discovered theory of gravity that unlike General Relativity would be consistent with quantum mechanics.

quantum mechanics: that part of physics dealing with the very small, usually at the atomic level and below, and characterized by concepts of matter and energy as simultaneously possessing particle and wavelike properties.

quantum physics: same as quantum mechanics, but an historic catchall that includes quantum theory precursors to quantum mechanics early in the century.

radioactivity: property of unstable atoms whereby they spontaneously emit radiation.

radioisotope: an atom that spontaneously emits radiation.

Rand Corporation: a private American corporation specializing in studies of military and geopolitical issues, the quintessential "think tank".

reactor: short for nuclear reactor, a device including a core and cooling system that contains uranium or plutonium arranged with control systems and a moderator to facilitate a chain reaction.

Safavid: a Persian dynasty that ruled Iran from the 16th to the 18th centuries, changed the country's religion from Sunnī to Shī'ite and gave it a national consciousness that eventually made it an important power.

SAGE: "Semi-Automatic Ground Environment", a U.S. air defense system designed to protect the country from attack by Soviet bombers.

Sanskrit: a classical literary language of India dating to the second millennium B.C. and still used in evolved form.

Sassanid: Persian dynasty founded 400 years before it was destroyed by the Arabs in the mid seventh century.

Seljuk: dynasty coming from Turkic tribes in central Asia ruling the central and eastern regions of Islām, particularly later in Anatolia as the Sultanate of Rum.

Sevener: name often given to the Ismā'īlī sect of Shī'ite Islām that split with the Twelvers over who should be the seventh *imām*.

shahādah: the first of the "five pillars" of Islām, the declaration of faith; "There is no god but Allāh, and Muhammad is his Prophet."

sharī'a: unchanging Muslim law, derived from God's word through the Qur'ān, governing all aspects of life; religious, secular, public, or private.

shī'ite: from the arabic, *shī'at*, party or follower (of Alī), those who believe that the successors to the Prophet should be of his family through his daughter, Fātima.

Social Democratic Party: Marxist Russian political party formed about turn of century and largely involved in supporting labor's rights.

Social Revolutionary Party: radical Russian political party formed about turn of the century largely supporting peasants and tending to terrorism.

soviet: a council to lead the people.

***starets* (pl. *startsy*):** a Russian monk who is a confessor and spiritual advisor.

Star Wars: informal name applied to Strategic Defense Initiative (SDI), a program to provide space defense against Soviet ICBMs.

Sūfī: of or pertaining to a Muslim mystic, Sufis were very effective missionaries for Islām.

sultan: "authority", used from the time of the Seljuks for the political head of state.

sunnī: one who believes in the *hadīth* and the consensus of the Muslim leadership and community; about 90 percent of Muslims are sunnī, about 10 percent shī'ite.

taqīyah: "caution," "dissimulation," "prudential concealment," a fundamental shī'ite tenant, the practice of concealing one's commitment to shī'ism; historically, often a necessary tactic for survival amidst the sunnī majority; also sanctioned and practiced by sunnīs when individually under serious threat by non-Muslims.

transuranic element: elements heavier that uranium, well over a dozen are known, for example, neptunium, plutonium, americium, and curium.

TNT: trinitrotoluene, a common high explosive the energy release of which is used as a standard of comparison for nuclear explosions.

Transoxania: the region between the Oxus and Jaxartes rivers.

Twelver: sect of shī'ite Islām that split with the Seveners over who should be the seventh *imām*. Their line of *imams* ended with the disappearance of the twelfth *imām* in the ninth century. Sect has strong messianic tradition awaiting the return of the Twelfth, or "hidden" *imām,* the *mahdī*.

Twelver shī'ism became the state religion of Persia under the Safavids.

ulamā: the class of religious scholars and leaders.

ultrasound: sound well above audible frequencies; used in medicine to observe the heart and fetuses with a radar-ke technique.

Umayyad: of or pertaining to the dynasty of Umayyah beginning with the first Umayyad caliph, Mu'āwiyah in 659.

umma: the community of all Muslims.

Uranus: the seventh planet from the sun, the next beyond Saturn.

Urgench: ancient city of Khwarizm about 100 miles south of the Aral Sea on the west bank of the Oxus river. Modern Urgench is about 80 miles southeast of the ancient city.

Venera: a very extensive series of unmanned Soviet probes to Venus between 1967 and 1983

Viking: U.S. unmanned probes to Mars.

Vizier: (or Wazier) chief minister.

Volta: Alessandro (1745-1827) Italian scientist invented one of the earliest forms of a battery.

Voyager: U.S. unmanned probes to Jupiter, Saturn (& Uranus and Neptune for Voyager 2.)

Wallachia: a region approximating the southern half of modern Romania.

X-rays: high energy radiation very much like violet light, but with a much higher frequency.

zakāt: alms, one of the "five pillars" of Islām.

***zemsky sobor*:** originally a group of prominent individuals appointed by the Tsar for a particular task, sometimes included even peasant representatives; last convened in mid 17th century.

Zond: unmanned Soviet lunar probes in preparation for subsequently cancelled manned Moon missions.

Zoroastrianism: pre-Islāmic religion of Persia founded in the sixth century B.C. has had important influence on Judaism, Christianity, and Islām. Largely destroyed in Persia by the Muslims, still exists in small groups, mostly in India.

For More Information

Islam, Arabs, and the Middle East

Esposito, John L., *Islam: The Straight Path,* 1991. A popular description of the beliefs, practices, and a little bit of history of Islām.

Hourani, Albert, *A History of the Arab Peoples,* 1991. A first rate account of the history of the Arab peoples as Arabs and not simply Muslims, right up to the present.

Ibn Khaldun, (trans. Franz Rosenthal) ed N. J. Dawood, *The Muqaddimah: An Introduction to History*, 1967. Not only does this work provide insight into the medieval Muslim mind, but it gives us an opportunity to learn from one of the great historians in all of history.

Lewis, Bernard, *The Political Language of Islam,* 1988. A nice little book useful to nonspecialists interested in understanding something about the Muslim political world view.

Lewis, Bernard, ed, *The World of Islam: Faith, People, Culture,* 1992. This huge paperback is a wonderfully illustrated, very broad and readable presentation of the title material.

Loeffler, Reinhold, *Islam in Practice: Religious Beliefs in a Persian Village,* 1988. This extremely revealing picture of popular Muslim belief before and after the Iranian revolution should be mandatory reading for all in the West in order to bring some rational understanding to the hysterical confrontation with Shī'ite Iran.

Saunders, J. J., *A History of Medieval Islam,* 1965. A good overview of the history of Islām up to the end of the Mongol conquests.

Sivan, Emmanuel, *Radical Islam: Medieval Theology and Modern Politics,* 1990. A useful and levelheaded presentation of the sources and beliefs of some Muslim fundamentalists.

Russia and the Soviet Union

Blum, Jerome, *Lord and Peasant in Russia from the Ninth to the Nineteenth Century,* 1964. A comprehensive survey of the evolution of serfdom from beginning to end.

Bobrick, Benson, *Fearful Majesty: The Life and Reign of Ivan the Terrible,* 1987. A good book to get a sense of Russia emerging from the isolation of the Tatar Yoke and yet still imprisoned by the terrible power of a questionably sane absolute monarch.

Cherniavsky, Michael, *Tsar and People: Studies in Russian Myths,* 1961. This book is very useful in trying to understand the Russian people and how they view themselves and the world.

Gilbert, Martin, *Atlas of Russian History,* 1972. This immensely helpful book is entirely of maps with just enough text on the map page to get one's bearings. Highly recommended.

Halperin, Charles, *Russia and the Golden Horde: The Mongol Impact on Medieval Russia History,* 1987. A short account and analysis of the complex Mongol impact, negative and positive, on Russian history.

Hingley, Ronald, *Russia: A Concise History,* 1991. Among the better short histories of Russia.

Massie, Robert K., *Peter the Great: His Life and World,* 1980. Everything you want to know about Peter and his Russia squeezed into 850 readable pages.

Massie, Suzanne, *Land of the Firebird: the Beauty of Old Russia,* 1980. The political turmoil in Russian history often obscures its cultural contributions. This book does a nice job of blending Russian history with Russian art.

Milner-Gulland, Robin, and Nikolai Dejevsky, *Cultural Atlas of Russia and the Soviet Union*, 1989. Like many "Facts On File" books a fine, balanced summary with first rate maps and illustrations.

Riasanovsky, Nicholas, *A History of Russia,* 1993. If one could have only one book on Russian history, this is the one. New editions continue to keep it up to date.

Tompkins, Stuart Ramsey, *The Russian Mind: From Peter the Great Through the Enlightenment,* 1953. Describes the intellectual currents that preceded the revolutionary era and thus are among the sources of our experience with 20th-century Russia.

Troyat, Henri, *Catherine the Great,* 1980. The last of the great quartet of Russian sovereigns that defined Russia apart from the peasantry.

Science and Technology

Aaseng, Nathan, *Twentieth Century Inventors,* 1991. Short biographies of 10 inventors including Robert Goddard (rocketry), Howard Aiken (computers), and William Shockley (the transistor.)

Davies, Paul, and John Gribbin, *The Matter Myth: Dramatic Discoveries That Challenge Our Understanding of Physical Reality,* 1992. This book by two very capable scientists turned science writer treats a wide range of physics subjects including cosmology. Included particularly for its charming and reassuring "Interlude" chapter where they "confess" to their own struggle to understand many of these weird subjects.

Eddington, Sir Arthur, *Space, Time and Gravitation: An Outline of the General Relativity Theory,* 1953. Written for the intelligent layman, this classic first published in 1920 remains one of the few works on the subject that gives a taste of the underlying mathematics without being a 'mathematical' book. Clearly written by a master.

Flamm, Kenneth, *Creating the Computer,* 1988. After a brief summary of the early history of computers this book focuses on post World War II developments and the government's decisive supporting role.

Gleick, James, *Chaos: Making a New Science,* 1987. This award winning book not only describes the discovery of chaos but presents extensive historical and biographical material on its discoverers. An extremely captivating 'must read.'

Morris, Richard, *Cosmic Questions,* 1993. A relatively short book that deals with thebirth, characteristics, and evolution of the universe in the form of questions with nice'bite sized' answers.

Murray, Bruce, *Journey into Space: The First Three Decades of Space Exploration,* 1989. A personal narrative by a former director of NASA's Jet Propulsion Laboratory. The focus is on American manned and unmanned space programs, though Soviet achievements are sufficiently noted to give a good sense of historical context.

Neal, Valerie, Cathleen Lewis, Frank Winter, *Spaceflight: A Smithsonian Guide,* 1995. Authoritative, nicely balanced presentation in all respects with lots of good photographs.

Reid, T.R., *The Chip: How Two Americans Invented the Microchip and Launched a Revolution,* 1984. An engaging story of the context and process involved in inventing and marketing one of the most important inventions ever.

Rhodes, Richard, *The Making of the Atomic Bomb,* 1986. A massive work of almost 800 pages suffering only from a tendency to over-write and include technical detail and barely relevant lengthy digressions. Otherwise as good or better than anything else on the subject. (Alas, his more recent work on the making of the hydrogen bomb, "Dark Sun..." has succumbed completely to the faults of the former with over 700 pages mostly on Russian spying. Not recommended.)

Walter, William J., *Space Age,* 1992. A nontechnical narrative of the history of the space age with lots of biographical information and colored pictures.

Index

Index

Index

Index

Index

Index

Index

Index

Errata:

Page 6, line 3: *For* epressions, *read* depressions

Page 56, line 36: *For* sewn, *read* sown.

Page 144, line 25: *For* principal, *read* principle

Page 152, line 20: *For* me, *read* and quickly became

Page 162, line 22: *For* essential, *read* essentially

Page 171, line 30: *For* trauma, *read* trauma of

Page 177, line 6: *For* odorous, *read* odious

Page 185, line 21: *For* them., *read* them. The reason was obvious even then.

Page 185, line 27: *For* drastically, *read* radically

Page 187, line 16: *For* is, *read* is no

Page 232, line 21: *For* being to be, *read* being

Page 237, line 3: *For* quantum, *read* foundation of the quantum

Page 237, line 9: *For* discovery of the quantum of radiation, *read* quantum hypothesis

Page 243, line 37: *For* This was a, *read* A

Page 243, line 38: *For* for only one atom to release, *read* per atom is released

Page 250, line 17: *For* 20, *read* 20,000

Page 251, line 36, *For* principals *read* principles

Page 257, line 18: *For* 30, *read* 30,000

Page 262, line 20: *For* Figure 6, *read* Figure 7, page 268

Page 262, line 20: *For* bottom, *read* top.

Page 326 line 41: *For* ibn Khaldun, *read* Ibn Khaldun (3), (23), (27), (46), (79), (80), (179)